Pieter Fischer

Music in Paintings
of the Low Countries
in the 16th and 17th Centuries

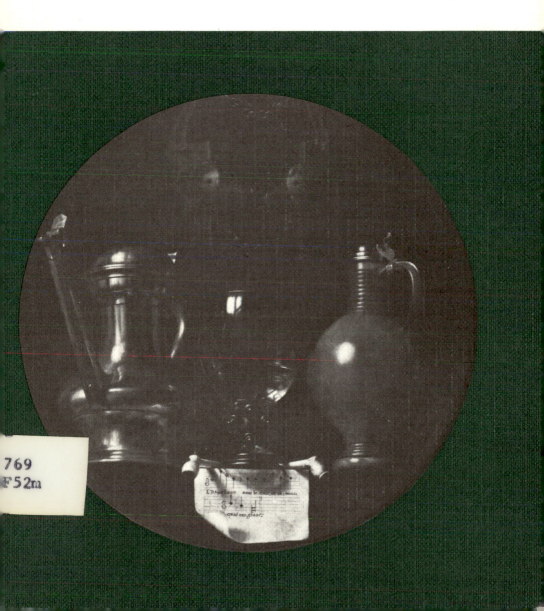

Music in Paintings
of the Low Countries
in the
16th and 17th Centuries

by

Pieter Fischer

SWETS & ZEITLINGER - AMSTERDAM
1975

Originally published as issue 50/51 of Sonorum Speculum by the Donemus Foundation of Amsterdam. Enlarged reprint by permission of the Foundation Donemus, Amsterdam.

ISBN 90 265 0185 4

Layout: Jack Jacobs

Cover: Johannes Torrentius, Still-life 1614 (Amsterdam, Rijksmuseum).

Printers: Erven E. van de Geer, Amsterdam (original edition) and
Miedema Press, Leeuwarden (enlarged reprint).

Contents

Preface

This slim volume is only a book as far as its form is concerned. It is actually the second printing, in a somewhat extended version, of a double issue of 'Sonorum Speculum', a magazine published by the Donemus Foundation, this particular issue marking the Foundation's 25th anniversary in 1972.

The new version obviously — and perhaps inconveniently — shows the traces of that origin. It halts between two ideas, an ambiguity for which there is naturally a reason: the hundreds of demands that kept on arriving after the jubilee issue was out of print. On the one hand we wanted to exploit the advantages of the bookform; on the other hand we did not want, for financial reasons, to extend the opusculum in the manner to which both the book-form and the subject gave full occasion. It is to be hoped that a book on a larger scale can be produced in due course.

A few remarks here about the subject and its treatment: the fundamental idea of the book came from Mr. E. Brautigam, former head of the information and propaganda department of the Donemus Foundation. In his opinion this type of subject, which is not usually treated in the magazine, but which must always have appealed to those interested in music all over the world — the frequent occurrence of music in the world-famous paintings of the Netherlands in the Golden Age — was the very thing for a jubilee issue. It was a favourable time, because of an increasing need to integrate the arts and finally expose their common context in people's lives. Music-lovers have long been interested in the visual arts, but almost exclusively as 'consumers' and in a 'utilitarian' manner: simply in order to increase their musical knowledge.

And this one-sided, narrow-minded mentality still unfortunately persists. Many people, however, want something better. On its own side, art history has been enriched for some decades by iconography, but if there is one other art-form apart from visual art whose nature is iconological, it is music. It is the aim of this book to demonstrate this, in connection with that other art, but also — with all due conciseness — with philosophical, sociological, aesthetic aspects. This resulted of necessity in a dual nature: occasionally details of the material are entered into more deeply than had previously been the case; on the other hand certain propositions and ideas are often presented which some readers might feel to be premisses worthy of more thorough clarification. Here are desiderata for the hoped-for larger-scale work. There are others, too.

The selected material calls for more 'environment', more context, such as: the connection and differences with art-expression in this area in other European countries, particularly Italy and France, the elaboration of certain themes in a broader context, colour illustrations and more details.

A final remark: I have intentionally returned some names familiar elsewhere to their original language. I wanted to free them from an atmosphere that is both rigidly scientifically 'justified' and unwarrantedly traditional, and to place them in their own reality perceived with awareness. International masters such as the 'Sadlers' **did** write their name according to a German mispronunciation, but were nonetheless called De Sadeleer (-laer); in Antwerp, Dutch 'Maarten' and French 'Martin' was called (and often wrote his name) Martijn de Vos — to say nothing of the name Brueghel.

I hope that this little book will provide an impetus for a continuation on a larger scale, and that it might help render our acquaintance with the treasures of our old art and music-making deeper and more real.

Preface to the First Edition

On the occasion of the 25th anniversary of the Donemus Foundation, the editors of Sonorum Speculum felt it appropriate to give this issue of the magazine a special character with regard to both the contents and presentation. We are privileged by the consent of Pieter Fischer, the Dutch musicologist and art historian, to let us publish his study on music of the Netherlands in the visual arts.

Owing to the length of this study it was decided to present this issue as a double publication of Sonorum Speculum. The importance of this subject, which is treated here for the first time to our knowledge, leads us to hope that this jubilee issue will be appreciated by our readers.

Dr. Jos Wouters, Chief Editor

The Author

Pieter Fischer was born in 1921 at Fort de Cock (Sumatra), and studied musicology and art history at Amsterdam University. He also studied organ with Anthon van der Horst at the Amsterdam Conservatory.

Pieter Fischer now lectures in music aesthetics at Amsterdam University, and teaches cultural and musical history at the Amsterdam Conservatory. He also conducts the Schola Cantorum Amstelodamense, which specialises in Gregorian plainsong.

English translation by Ruth Koenig.

Introduction

Musical notation — fascinating, not just as
a promise which is fulfilled on Hearing,
but a fascinating Sight, too:
as a play of lines and dots,
obeying their own laws and relationships,
grown through the centuries of craftsmanship
into writing, like handwriting,
just as Bach's notation appears to us; or as
Dusi, restorer of paintings, in 1839 completely 'rewrote' and 'characterised' them in a
virtuoso manner on a Conversazione Sacra
by Carpaccio, painted in 1516 for the cathedral of Capo d'Istria[1]), and as the designer
chose Willem Pijper's notes for the cover
of 'Sonorum Speculum',
making them no longer mere 'sonorum sig-
na' — sound signs — and something more
than 'sonorum speculum' — sound mirror:
an abstract visual work of art and an image
of themselves, abstracted even from notation's actual task as a symbol of the art of
sound, —
not until the twentieth century have they
really been appreciated.

And yet musical notes appeared in great
number in works of art before this century:
from the middle of the fifteenth century in
fact, with uninterrupted regularity[2]). Since
critical eyes were never aesthetically offended by their presence, we have to establish
the fact that musical notation never occurred in visual art just in their function of serv-
ing non-visual art, as 'corpus alienum', as
a 'foreign body'. Not only did the painters
and sculptors understand the art of placing
them in pictorial harmony with their painted
surroundings, they were also able to endow
the notes with important symbolic functions
of diverse kinds, from the simplest to the
most complex and exalted.
Even when musical notes retained an important part of their musical, non-visual function, in portraits of composers and musicians, for instance, we see them from the
very outset in emblematic and symbolic
values, which we can sum up in various
ascending steps:
— as *general attribute:* 'this is a musician',
just as a man with hammer and saw is seen
to be a carpenter; it is not necessary to
know *what* music it is; the artists found
various means of expressing this, such as
painting signs which merely suggest notes,
or 'just any notes' with no musical connection;
— as a *specifically determining attribute:*
just as an Apostle depicted with two keys
is obviously Saint Peter, or as a young woman playing the organ and singing with angels denotes Saint Cecilia, a particular fragment of music or short piece will announce
its maker: 'this master established his fame
with this piece', or 'he excelled in this type
of work'. A group of composers' portraits
painted between 1575 and about 1650 thus
show musical canons, the contrapuntal feat
which was regarded as the height of musical craftsmanship at that time, possessing
the force of an emblem and a signature. 'Of
course you can recognise the subject by
this canon: it is Master...'. This type of
portrait chiefly flourished in Germany; portraits of Samuel Scheidt and Michael Prae-
torius in their works 'Tabulatura Nova' and
'Syntagma Musicum' are examples. The por-
trayal of composers with a sheet of music
did not become common in the Netherlands
until the eighteenth century, when it appeared as a result of foreign influence and
as a part of a general European phenomenon;
— not merely as a *general* or *particular*
attribute, but as a real *symbol* now: in 1761
Louis-François Roubilliac chiselled at the

orders of the chapter of Westminster Abbey in London the words and music of 'I know that my Redeemer liveth' on Handel's sepulchral monument; the notes do not only refer to the deceased's occupation and to a famous aria from a famous work, 'The Messiah', but the choice of this piece of music indicates Handel's belief in life and redemption after death. The word 'I', in the 'Messiah' meaning man in general, and Handel's own music acquire a more acute and significant meaning[3]), especially since his death coincided with that of his Redeemer, during the night of Good Friday, April 13th-14th 1759, in expectation of the Resurrection at Easter. In the 'Gazetteer and London Daily Advertiser' of April 16th, a poem appeared which included the following words:

'Ah! when he late attuned Messiah's
praise . . .
Messiah heard his voice — and
Handel dy'd.'

The lack of a portrait of Dutch musicians with notation is amply compensated by the abundance of notes that occur in Dutch paintings, especially from the sixteenth and seventeenth centuries, painted in open books or on sheets of music, not connected with the composers, but with other people or objects.

Freed from its too-functional relationship to the composer, painted notation can not merely assume the above-mentioned stages of attributive and symbolic significance, but can transcend them by far: in Dutch painting it forms the reflection and compendium of the most widely differing opinions about music as they have been formed and developed in European culture since the periods of antiquity and early Christianity. This aspect will be the main subject of this essay. If our only aim here were to make as complete a list as possible of identifiable music that can be recognised in paintings, so as to satisfy a materialistic thirst for knowledge we should not only be impoverishing the area — moreover the amount that can be recognised is not so big! — but should also be letting the real meaning, the essence and most fascinating aspects of both painting and music elude us because of our disregard of the intellectual backgrounds and interrelations of both art-forms. Since the revival of interest in old Dutch music in the nineteenth century and up to the second World War, connoisseurs and lovers of music have used visual art in order to wrest from it knowledge about people, music and the way it was played; this was necessary but was merely the surface, the skin. The revival of interest in old Dutch painting was parallel: museums were built industriously in a period of intellectual poverty and materialism, at the nadir of understanding for old art — when it was a virtue to throw overboard the ballast of allegory, emblem allusion and context, retaining only a purely aesthetic-materialistic attitude and raising i to become the only and highest ideal, corresponding to the idea of the period: art for art's sake, fruit of romanticism. A still-life by Teniers in the Brussels Museum, charged with significance, still bears the title which it was given during those years: 'Etude d'accessoires'. Those responsible for this title did not realise that the word 'accessoires' had a faint aftertaste of the old meaning: 'requisites', of a moral lesson in painting — what was meant at the time was merely 'incidentals', an arbitrary collection stripped of meaning, of unimportant, faded crinkled articles.

Today, feeling for the emblem has reawakened, iconography has taken to its heels. We are capable once more to a certain extent of conjuring up and imagining the spirit of the time in which these old works were painted: how the spirit of the contemporaries of the Brueghels, of Rubens, Rembrandt and Vermeer were nourished, excited and inspired, especially by an intellectual-aesthetic wealth, taking precedence over pleasure in superficial aesthetics. However, iconography and emblematics are also extremely dangerous: in contrast to fashion and the undisciplined flights of fancy which parasitise the new goldmine, Prudence and Temperance can show us the way to a world of real values and pleasure.

However, it was neither painting nor music that was the chief domain of symbolism: life itself had from earliest times been regarded as charged with many hidden meanings; in this respect, our entire occidental culture remained throughout the centuries in the power of the Antique world and early Christian vision of life, both of which in their turn depended on even older eastern views. The arts were part and expression of that life itself, and as such bearers of the symbolic values; music to a greater extent, really, than painting, the latter being more an expression, a report of life, and music the expression *of* life *in* life, in which it had taken the very first place since oldest times.

Remnants were preserved and revived in the music of the Renaissance and Humanism, and thence in musical opinion and practice of the sixteenth and seventeenth centuries. Painting has illuminated an astonishingly visual picture of this, an intensified picture, framed as it were, as though seen from a dark room (Camera obscura), like a lantern slide. Paintings from that time thus help us towards a deeper insight into music, whilst music and painting comment on each other for a better understanding of life's hidden meanings. There is no sufficient space here to go into the development of the concepts of 'meanings', functions and roles of music in life since ancient times. For this, readers are referred to the 'family tree' at the end of this article. These variants will also be mentioned with reference to the paintings under discussion.

The Netherlands occupy a special place in the field of music in painting among the European nations: together with Italy they head the ranks in number and quality. In spite of confining ourselves to Dutch art, we also have an opportunity of demonstrating how both art-forms have been mutually fruitful, although our selection will mainly show the Italian influence on Dutch art.

„Tempus et Aeternum"

'Tempus et aeternum: haec duo nostra
vocantur;
haec meditanda homini — caetera mortis
erunt'.

('Time and eternity: these two things can
be said to be ours; on these things must
man meditate — the rest are for death.')
This distich from a poem by Jan Albert Ban,
which will have such an important part in
the development of the common paths of
music and painting in the Netherlands, may
serve as 'Leitmotiv' in our discussion.
'Time and Eternity': they give a precise indication of both the separation and the link
between the two Netherlands, their culture
with their music:
the Catholic South of the latter half of the
sixteenth century and the beginning of the
seventeenth, the South that with its Gothic
tradition formed an especially fascinating
and fertile alliance with the humanism which
was guided by the counter-reformation, the
South that regarded music chiefly as the
perfect image and even the realisation of
the Praise of God, the final goal of all earthly things: heavenly eternity; — music, then,
as the symbol and essence of *Eternity*, the
'gloria in excelsis' which the Angels have
always sung, will be the highest song into
all eternity;
the Protestant North of the seventeenth
century, responsible for an equally interesting form of culture: Humanism introduced
into the Dutch Calvinist way of life, in
which music is specially created as the
image of what is temporal, transient, vanish-
ing, fading away — music as the symbol
and essence of what is *Temporal*. There is
no clearer example of this than J. P. Sweelinck's 'Vanitas Vanitatum' (1608).
It is remarkable how the North was nourished by the South. The main point of encounter was Jan Albert Ban's Haarlem, the Catholic and 'Antwerp-Protestant' Haarlem:
the most interesting Antwerp Intellectuals
spent shorter or longer periods there on
their way to the North, for in Haarlem there
was a wealth of intellect between 1580 and
1620 that was unparallelled elsewhere: at
that time Amsterdam depended more on
Haarlem, it was still rising, but was
forming together with Haarlem the great *axis*
of development; Utrecht and especially Leiden were to second them in importance. The
North, however, was not a mere dependent
of the South; on the contrary, it provided its
own, typical interpretation of the rich material from Antwerp.

The South
Netherlands 1565-1620

The Praise of God and the Good Earthly Music

The Praise of God

The surprisingly large amount of musical compositions in Dutch paintings in the second half of the sixteenth century was not a new phenomenon, but a continuation of a tradition whose roots reach far back into the Gothic period. First musical *Instruments* occurred in increasing numbers in sculptures, carvings and manuscripts from the twelfth century; not until after 1400 did musical compositions appear. One fact is plain: in the Middle Ages there was officially only one true form of music — sacred music, as a prelude to and premature participation in celestial singing. In this theocentric culture, profane music still bore the odium of fiddlers and jugglers, music inspired by the devil and temptation; fiddlers and other folk-music instruments did appear in paintings, but mostly to a sarcastic or moralising purport. The contrast to church music was extreme, the latter being the Divine music which ancient church tradition insisted would be sung to God by the nine choirs of angels, world without end. The miniatures with which Dutch painting started showed an increasing tendency to depict detail with great clarity so that the eye might be deceived into thinking that it could see the actual object: 'trompe-l'oeil' had arrived. But musical notes are too small for this, and the first attempts to bring music to a new life in paintings were not seen until the end of the miniature's supreme reign, just before the rise of oil-paintings on panels, when, the miniature became increasingly large.

It is not surprising that these notes appear as hymns of angels singing the glory of God in Bethlehem, descending from the invisible celestial halls to the earthly vaults of heaven in order to continue the Song of Praise for God incarnate: in the famous 'Très Riches Heures du Duc Jean de Berry', illuminated between 1413 and 1416 by the Nimwegen brothers Paul, Jan and Herman van Limburg, we see these Angels singing from sheets of music twice: at the 'Glad Tidings to the shepherds at the Birth of Christ'[4]) and at the 'Adoration of the Magi'[5]). The notes themselves do not answer the purpose of an existing or singable music, but the red letter G on the second miniature leaves us in no doubt as to what the Angels are singing: 'Gloria in Excelsis Deo', as written in the Gospel. This 'Gloria', with other similar texts to the glory of God or the Holy Trinity, was to remain the root of reproductions of celestial music in Christian art.

As far as we know, painters of the South and North Netherlands included the first musical compositions in their paintings during the latter half of the fifteenth century. We might have expected them to be by Robert Campin, by the master of the 'Annunciation of Aix' or by Jan van Eyck in the first half of the century — by those sharp observers of realistic detail, the heirs of the miniature painters. But we do not know of any. In Jan van Eyck's 'Lamb of God' in the Church of Saint Bavo at Ghent, the Angels sing God's Praise from a book, the back of which is turned towards us; the organist Angel has no music in front of him. The singing Angels in 'The Source of Life' do not transcend the miniature tradition in this respect: mock notes and scrolls of text[6]). May we think of lost works such as 'Saint Hieronymus in the study' which Antonello de Messina continued, and Carpaccio, who depicted music in his 'Vision of

Saint Augustine in his study'?

Be that as it may, the oldest Dutch work of art we know which as a musical composition on it is Hugo van der Goes' Holy Trinity altar for Sir Edward Bonkill, probably from 1473. Like van Eyck's 'Lamb of God', there is an organ-playing Angel, but this time with an open book of music in which the hymn 'O Lux beata Trinitas' can be seen in Gregorian notes[7]), a clear testimony of the typically English devotion to the Trinity. A little later there were compositions in paintings which in their turn, just as so many other religious productions of the time, testify to a revival of devotion to Mary. In this connection I refer to the little panel which is an early copy of a painting by Geertgen tot Sint Jans of Mary as queen of the Rosary, surrounded by concentric circles of choirs of angels with diverse musical instruments. Many can also be found on paintings in which her praise is sung with painted notes in works by several, unfortunately mostly anonymous, but fine masters around 1500: particularly the Master of the Embroidered Foliage with the motet 'Ave Regina Coelorum, Mater Regis Angelorum' by the English composer Walter Frye, who worked in the South Netherlands and at the Burgundian court[8]), the painting illustrating this hymn: Mary enthroned with the infant Jesus, surrounded by angels playing musical instruments, one holding the open book showing the motet, whilst two more angels hover above Mary, holding a crown, — the Master of the Saint Lucia Legend, with the same text but another composition, on a panel representing the Assumption and Coronation of Mary, [8a]), — and the anonymous painter identified by Van Regteren Altena as the Master of Saint Aegidius (Saint-Gilles), possibly Jean Haye, who depicted a similar representation on the centre panel of a tryptich made for Louis XII of France.[9])

At this point a remark must be inserted about Hieronymus Bosch, in whose work music occupied an important part. The notes he painted always were those of profane music inspired by devils. Bosch never made any efforts to reproduce a real composition of such a deplorable nature; the notes are mere indications. It was not Bosch's intention, nor did it suit his pictorial style, to dissect the numerous large and small requisites necessary for the momentum of his comprehensive visions and to distract attention from them. The detail in his work is meant each time to reveal more strangely the great events with which he is concerned, and his fine, sometimes sarcastically fine brush never performs 'Feinmalerei' as it was called later. We only find real notes in later replicas of his work. The most famous example is the so-called 'Concert in the Egg', an anonymous copy from just after 1560 of a lost Bosch original; the painter of the copy painted a recent polyphonous song on the song-book of the singing seedy figures, with a significant text: 'Toutes les nuicts que sans vous je me couche . . .' ('Every night that I sleep without you . . .'), after the composition by Crecquillon from the 'Septième livre de chansons à 4 parties', which was published at Louvain in 1560 by Phalèse[10]). But the real spirit of Bosch is no longer present in such replicas, one of the reasons being this very concretisation of the music. They exhibit a newer mentality — we shall refer to this again later. Bosch was however the great exception of his time, not having music's role as Praise of God in mind with his over-subtle criticism and sarcasm, but having in mind the very opposite, as in all his other work.

Master of the Saint Lucy Legend, Mary, Queen of
Heaven (Washington National Gallery of Art)

Mary, Queen of Heaven (detail)

Pieter Huysch, after Hieronymus Bosch, Concert
in the Egg (Musée de Lille, Photo Giraudon)

The sixteenth century brought great changes to Europe. A prime one, besides the discovery of the world and religious reform with the counter-reformation, was the large-scale and definitive extension of the Italian late Renaissance and of Italian Mannerism over Europe, the originally Florentine humanism being one of the chief factors. The South Netherlands greeted this new cultural impulse enthusiastically, but united it with its own late-Gothic tradition, resulting in a typical South Netherlands version of Mannerism, Humanism and Early Baroque. Having found a willing ear in the old Gothic symbolism that still existed, Italian humanistic emblematism, itself linked with ancient emblematism and mythology, began to prevail.

Music had its place here too. For some time it looked as though this conception of music would occupy a dominating position at the cost of the Gothic Praise of God. A print published by Hieronymous Cock at Antwerp in 1554 after Maerten van Heemskerck[11]) demonstrates one of the chief characteristics of the humanist conception of music. This characteristic is the moralising tendency, a reference to the good and evil in everything, an admonishing finger. This somewhat strict, mistrustful, occasionally pessimistic view of life is the heritage of ancient Greece, where Plato's voice was loudest, certainly with regard to music too: there is good *and* (take special note!) bad music[12]). Music in itself is good, but can lead all too easily to evil: humanistic pessimism even regarded music as the main source of evil, of moral retrogression. On the said print people can be seen in the foreground, people who ought to know better, clerics with drunken expressions pursuing a satyr playing his pan's pipes with an open song-book on his back showing a

four-part piece which the pursuers are trying to sing. The notes are clearly visible but do not form a recognisable piece of music. Music here is one of the lamentable sequels of Bacchus' triumphal chariot. Similar attitudes are propagated by masters such as Lambert Lombard, his pupil Frans Floris and the publisher of prints of their work and that of others, Hieronymous Cock, and also Pieter Pourbus.

But the tide turned around 1560. The humanistic conception of music — to which we shall return — was not lost, but had to relinquish its domination again to the 'Praise of God'. This was due to the thrust of the Counter-Reformation. Under the leadership of the Jesuits, especially in the artistic metropolis of Antwerp, the South Netherlands counter-reformation was to take up a thread of the old Gothic tradition again, and use it to create a new style of 'God's Praise' — one of the main subjects of this essay. This new religious spirit was also to have a fairly strong effect on the humanistic conception of music that still prevailed in secular music, as we shall see in several cases later.

It was probably the influence of Roman painters that Michael Coxcie, who lived in the Mechelen of the Archdukes, and Jan Gossaert, who had also lived there, painted Saint Cecilia, the patron saint of (celestial!) music performing an existing composition together with angels. He made a larger version of this painting, which Philip II of Spain commissioned in 1569, and a simpler one, a replica of which is owned by Professor K. Ph. Bernet Kempers — appropriately so, since in the painting the upper part of the motet 'Caecilia Virgo' by Jacobus Clemens non Papa is shown, a composer whose works have been edited by the owner of the painting, who had previously written his

Michiel Coxcie, S. Cecilia, (Madrid, Prado)

Michiel Coxcie, S. Cecilia, detail

Michiel Coxcie, S. Cecilia, detail

doctoral thesis on this master.

The larger painting in the Prado is more elaborate. The young angel in the left-hand lower corner has a soprano part open at the same motet, 'Caecilia Virgo Gloriosa'. Although the 'secunda pars' ('Biduanis et triduanis jejuniis orans' . . . etc.) of this motet is evidently to be found on the next page of the book and therefore invisible to us, the painter wished to show us this important text and painted it on the left-hand page before the beginning. Twofold clarification is necessary for the tenor part, which is at Cecilia's side on the spinet. To begin with, the painted music is not the tenor part of the motet 'Gloriosa', but of another motet from the same collection: 'In lectulo meo per noctes quaesivi quem diligit anima mea . . . et non inveni' ('at night I sought he whom I love in my bed . . . and found him not') from the Song of Solomon. The painter chose this text intentionally, since it symbolises both Cecilia's virginity and her marriage to Valerianus: an official marriage in the eyes of the world, but only a sham, since she also converted Valerianus to Christian virginity. However, the notes painted here are not from the tenor part (as at the top of the page) but from the soprano part (the key-signature can just be seen at the beginning of the line). Perhaps an assistant of Coxcie's made a mistake? Or is this a symbol too — the voice of *Cecilia*, in whose mouth the words are put?

The painting proves that Coxcie used an edition by Phalèse of motets by Clemens non Papa: 'Liber Quartus Cantionum Sacrarum vulgo Moteta vocant quatuor vocum . . . Lovanii, Ex Typographia **Petri** Phalesii' 1559, or a later edition, perhaps the second from 1562.

This Saint Cecilia is herself a proof of the source of inspiration: the Roman counter-reformation.

Not long afterwards this motive of the counter-reformation was linked in a charming manner with the old Gothic 'Praise-of-God' native tradition, which thus underwent a revival and a special development.

Whoever's idea it was, that of Martijn de Vos, the painter, or of Johan I Sadeleer (Sadeler, Sadler), the engraver (probably the latter[13])), a print appeared in 1584 representing a new interpretation of 'Sint-Anna-te-drieën', Our Lady and the Child with Saint Anne (hereafter referred to as 'the Saint Anne 'triad'', a typical phenomenon in the devotion and art of the Low Countries). Mary is shown as the adult Mother of God, and is plainly the chief figure, as confirmed by the following: above her is the complete four-part composition by the Antwerp master, Cornelius Verdonck, on 'Ave, Gratia Plena'[14]). This composition was only 'printed' in this manner, thus being 'the earliest known composition printed by engraving'[15]). Engraving flourished greatly during this period, and engravers searched ceaselessly for motives and series of motives, often seeking inspiration in older representations of tried popularity, as we have seen in the case of Bosch's works. If some aspects were too oldfashioned, they were adapted to the more modern ideas. Sadeleer, de Vos and Verdonck were surely familiar with the late Gothic statues and paintings in churches (executed less than sixty years back!), with the Saint Anne 'triad' as one of the favourite motives, especially in wood-carving, and also with paintings representing the Praise of Mary, altars and devotional pieces such as the one by the Master of the Embroidered Foliage with Walter Frye's motet, 'Ave Regina Coelorum'. These old motives were all revived here.

Martijn de Vos and Jan Sadeleer, S. Anne, Our
Lady and the Child, engraving (Ex. Den Haag, Gemeente Museum)

The old 'mystery', complete in itself, of the Saint-Anne 'triad' representation, drawing as it does the spectator's mind to the miraculous fact of two generations, which by divine intercession had been given a more immaculate birth than other humans, was freed from this isolated position and from its connection with the Saint's festival (July 26th), and ushered into the Christmas mystery, to which the text from Isiah alludes which can be seen on Anne's lap: 'Unto us a Child is born'. It became a devotional work for the Mother of God on Christmas Day. To the left of the throne something of the 'earthly life' of Mother and Child can be seen: Joseph working with some wood; at the right there is the place where Jesus 'ought to be' ('in his quae Patris mei sunt oportet me esse' — Luke II, 49), and where Verdonck's motet should be sung: the House of God. Furthermore: the mediaeval interpretation of Mary, who, although the Mother of the infant Jesus, in the St.-Anne 'triad' motive was represented as a little girl, offended the humanists; Mary has become a blooming young mother. This version also fitted in better with the emblematic series of the ages of man (still with the moralising background): childhood, adulthood, old age.

Seven cherubim descend in a cloud to the canopy under which the three figures are seated; three of the cherubim hold the book, and the other four perform the motet vocally and instrumentally. The ground is strewed with various sorts of plucked flowers, a tribute to Mary but also her symbols from ancient times: the 'mystic rose' in several kinds, the iris or 'sword-lily' and heart's-ease, indicating Mary's pain, lily-of-the-valley for her heavenly innocence, the carnation for immortality and eternal life. The text on the plinth below, from Luke VIII:

'My mother and my brethren are these which hear the word of God, and do it', is an equally strong link for all contemplative Christians with the infant Jesus, the Redeemer, and with his mother and kinsmen including his grandmother Anne and showing Mary's role as mediator in the Redemption. But the words 'which hear the word of God' also have a connection with music. This is another example of the linking of a humanistic emblem with the counter-reformational idea. 'The Five Senses' formed an extremely familiar emblematic motive; 'Hearing', especially, was connected with music. A painting which expressed 'hearing' showed musical instruments, singing people, music-books. Since the counter reformation, ancient morals had been placed on a 'higher' plane, in the field of Evangelic virtuousness. The text of Luke VIII, 21 and the similar one in Luke XI, 28 ('Blessed are they that hear the word of God, and keep it') stay connected with music in the category of 'hearing'.

This special and detailed print, although we do not know whether it was made with a direct motive, must have made a great impression: in the years that followed several such prints appeared, so that this Saint Anne 'triad' was the first of a type and of a series. In the next year, 1585, a print appeared of Mary's 'Magnificat' by the same three artists: Martijn de Vos, who painted it (the original has been preserved and is in the Catholic Church at Kannstadt (Württemberg)), Johannes Sadeleer, the engraver and Corneille Verdonck, who composed the five part 'Magnificat'. The print bears the mention that it was published at Antwerp during the extremely oppressive siege (by Alexander Farnese), 'in lucem editum obsessa arctissime Antverpia 1585', i.e., before the surrender of the city on the 17th of August

18

Martijn de Vos and Jan Sadeleer, Magnificat, engraving (Ex. Den Haag, Gemeente Museum)

Sadeleer's print of Martijn de Vos' 'Saint Cecilia' probably dates from 1586. Other composers come into the picture now. The young canon Daniel Raymond of Liège (Raymundi', ca. 1560—1634[16])) composed a five-part motet for Sadeleer on the text from the 'Passio' and the Office for Saint Cecilia, 'Fiat cor meum immaculatum'[17]), which was often used by Roman painters such as Domenichino[18]).

Sadeleer and De Vos also collaborated in 'The Song of Solomon', a print dated by Sadeleer as 1590. He had already left Antwerp three years previously and now lived in Munich, as we shall see further on. Perhaps he made the print in Antwerp, and completed and published it in Munich. The composer of the motet 'Osculetur me osculo oris sui' (the beginning of Solomon's Song) was another Antwerp master of this time: Andries Pevernage[19]).

Martijn de Vos supplied Sadeleer once more with a 'figuration' to be engraved: The Angels' Hymn of Praise, in which De Vos' italian (Venetian) background and the early Baroque of the counter reformation can be plainly perceived. Pevernage supplied the composition again, a nine-part setting of 'Gloria in Excelsis Deo' this time, continuing the old Netherlands tradition. The print is dated 1587 and was published in Mainz; this assigns the cause of the discontinued association of Sadeleer and De Vos: the famous engraver, his brother Raphael and young cousin Aegidius Sadeleer left Antwerp that year, never to return. The journey was to end in Italy, but took them to Cologne and Mainz and thence to Frankfurt, where Sadeleer stayed for three years, and where so many South Netherlands artists had established themselves, most of them for religious reasons, especially since the taking of Antwerp by Farnese in 1585.

Important engravers, publishers and painters were among them, from Antwerp, Brussels, Liège and other towns of the South Netherlands. The fact that Sadeleer himself was at first the only engraver to carry on working on the picture-motet, each time with another painter demonstrates that he was the soul and creator of this genre.

At Frankfurt Sadeleer found a new 'figurator' for his engravings in the genre in question: Joost van Winghe, born like himself in Brussels, and in Frankfurt since 1585, after having been court painter to the same Duke Farnese that conquered Antwerp in 1585. Two works have been preserved from their collaboration during Sadeleer's stay in Frankfurt:

The 'Adoration of the Lamb' from 1585, for which Andries Pevernage supplied a four-part setting of the Apocalyptic 'Dignus es Domine', a composition which bears a slight resemblance to the older famous Flemish work, Van Eyck's altar-piece at Ghent, but which is entirely translated into the language of the Italian counter-reformational early baroque[20]);

'David Singing God's Praise', the music to which was again supplied by Pevernage — with a typically counter-reformational text 'Laude pia Dominum magni vehat orbita mundi', set for five parts, the quinta vox proceeding from the Cantus by following the canon indication: 'Duo in carne una'; David is playing the harp from a book which might possibly be a sort of 'intavolatura' of the motet; the bed behind him symbolises his early rising to sing God's praise, which he continues to do until night, according to his own psalm: 'Septies in die laudem dix tibi' ('seven times a day do I praise thee (Psalms 119)), the original Van Winghe painting or a copy of it being in the Episcopal Museum at Haarlem.[21])

Joost van Winghe and Jan Sadeleer, David Singing
God's Praise, engraving (Ex. Den Haag, Gemeente Museum)

Pieter de Witte, David Singing God's Praise (Haarlem, Frans Hals Museum). Photo A. Dingjan.

Formerly Rügenwalde (Darlowa), Christoph Lencker,
silver relief after Pieter de Witte - Jan Sadeleer,
David Singing God's Praise

The Sadeleers went to Munich around 1590-1591, where Jan became court engraver. He devoted himself once more to 'his' genre, the picture-motet. He not only found his 'figurator' at the court of Munich — the Flemish court painter Pieter de Witte who called himself Pietro Candido or Petrus Candidus, but a new South Netherlands composer too — the court Kapellmeister and the greatest master of the Netherlands school of his time, Lassus, who called himself Orlando (di) Lasso.

The subject was again 'David Singing God's Praise', this time with a humanistic-counter-reformational variant of the 'instrumental psalm' 150. 'Laudent Deum Cithara'. A study by De Witte is in the Rijksmuseum at Amsterdam, [21a]) another, fragmentary, in the library at the University of Erlangen, the original painting which was Sadeleer's model being in the Frans Hals Museum at Haarlem. It only shows the Cantus and the Bassus, whereas Sadeleer's engraving contains the entire four-part composition. It is also striking that De Witte sketched an organ-playing angel (at Erlangen), to replace it in the painting by Saint Cecilia, wearing a crown: both patrons of church music thus being shown as King and Queen of the celestial music. Sadeleer followed the painter's example. He dedicated the print to Princess Maria Anna, the daughter of Duke Wilhelm V of Bavaria[22]). He also dedicated to Lassus, the composer, his engraved portrait, 1592[23]).

The engraving had an unexpected sequel. Philip II, Duke of Pomerania, ordered an altar to be made for his court chapel at Stettin in ebony with silver relief work; he chose Hendrick Goltzius' etching of the Passion of Christ as model for the reliefs. These were chased in silver by the Stettin goldsmith, Johann Körver, and after his death in 1607 by the Augsburg silversmiths, Christoph and Zacharias Lencker, father and son. The crowning piece came in 1612, and was a 'Himmlisches Conzert', executed by Christoph Lencker (the son died in 1612) using Sadeleer's Munich engraving. In 1636 the widow of the last Duke of Pomerania, Bogislav XIV, had the reliefs placed in a new ebony sculpture in the church at Rügenwalde, which is now part of Poland and is called Darlowa. The 'Silver Altar' disappeared without a trace after the second world war[24]).

As for Sadeleer himself, his picture motet seems to have arrived here at its end; his example however found some followers.

Sadeleer went to Italy in 1592, was heard of in Verona in 1595, and died at Venice. He doubtlessly visited Rome, where Netherlands artists all stayed for a shorter or longer period just then, many of them engravers from Antwerp or Haarlem. In connection with developments which will be broached later on in this essay, some names are worthy of mention: Goltzius, who knew Sadeleer very well, and was at Rome in 1590 and 1591; his stepson Jacob Matham, who was there between 1593 and 1597 — moreover, these were the years during which the famous Caravaggio created such a sensation in Rome.

The French engraver Philippe Thomassin, who worked in Rome, indubitably followed Sadeleer's example in his picture motet after Agostino Ciampelli's painting 'The Vision of Saint Gregory' with the four-part motet 'Regina Coeli Laetare' by Palestrina's pupil Francesco Soriano (Suriano)[25]). Seiffert connects the print with the Editio Medicaea of the Gregorian Masses of 1614-1615, on which Soriano and Anerio worked[26]). The picture itself points in a different direction; indeed, the inscription at the bottom, 'ad-

Agostino Ciampelli and Philippe Thomassin, Saint
Gregory the Great, engraving (ex. Biblioth. Nat.
Paris)

modum R.P.F. Gregorio Donato Romano, Ord. Praedic.', referred to the papal censor, who had to be put in a good mood by this picture of his patron saint with a motet by the influential and extremely orthodox kapellmeister of Saint Peter's, Don Francesco Soriano, the idea being that he should approve of the prints which Thomassin intended to publish in 1618, which were of ancient statues with a great deal of nudity[27]). The engraving and the motet must therefore date from 1617 or thereabouts. Soriano included the motet 'Regina Coeli' in his big edition of religious works in 1619: 'Passio D.N. Jesu Christi . . . Magnificat . . . et cetera'.

Philippe Thomassin (1562-1622) could have seen Sadeleer's prints with motets at Rome, either just before or in 1617. We should remember, though, that Thomassin had been very familiar with Sadeleer's work for a long time: he was one of the chief masters of the flourishing engraving school at Nancy, the town where he was born and where the Flemish art of engraving with its typical, manneristic style was the principle model, as can still be seen in the work of the younger generation of Nancy: as will as Callot (1592-1635) particularly Bellange (1594-1638).

The genre even lived on in several works by Domenichino during his Roman period (before 1630) — 'Saint Cecilia' in the Louvre, for example, and by Poussin — 'Saint Cecilia' in the Prado, and others.

There are two more interesting prints among the followers of Sadeleer's creative example, both coming forth from his immediate Netherlands surroundings.

Around 1590-1595, the engraver Philips Galle, who came from Haarlem, published his 'Encomium Musices' ('Praise of Music') at Antwerp, with a frontispiece based on a painting of the Antwerp painter, Johannes Stradanus, who worked in Florence. The painting shows a complete six-part song, again by Andries Pevernage, 'Nata et grata polo'[28]). At a first look at the title, print and music, this would seem to be a profane work, but it is really a fine example of humanism in the South Netherlands at the service of the counter-reformational revival of the 'Praise of God' tradition in painting and music.

The text of Pevernage's setting is a typically humanistic verse of two couplets:

'Nata et grata polo vocum discordia

concors,

Musica, scitque homines, flectere scitque

Deos,

Flectere scitque feras, at quisquis nescius

illae,

Flect (. . .), is nec homo, nec fera sed

lapis est.'

'Music is the creation and the will of the

heavens,

The concordance of divers voices,

It can move man, God, and beast,

But he who knows it not and is unbending,

He is neither man nor beast, but a stone'.

The illustration is a commentary on the first two lines: Musica, holding the book in which the music is written, is flanked by her sisters and the paranymphs Harmonia, who bears a winged heart and two ears (the Concors Consonantia and Harmony winging throughout the spheres) and Mensura (Measure, but also: the virtue of moderation), with a measuring instrument; they stand in the midst of a lavish collection of multifarious musical instruments, all sounding in one organised whole — vocum discordia concors Musica.

This entire humanistic representation, built up of musical and emblematic elements from ancient times and increased during the course of time, however, is completely at the service of religious music, the Praise of God: under the print we again see David's 'instrumental' psalm, Laudate . . ., Psalm 150, and after the frontispiece there is a series of sixteen proverbs with illustrations which Galle, as the title states: 'sacris litteris concinnebat' ('adapted from the Holy Scriptures'). Pevernage's composition is therefore, in spite of its profane and antique appearance, a motet.

The last picture motet which we shall discuss leads us to Leiden. A man lived there between 1595 and 1600 who was one of the most inventive and inspiring minds of the Netherlands, a man whose varied life and many contacts accentuated the links between the South and North Netherlands: Jacques de Gheyn (II). Flanders had produced a striking number of such creative intellects during the sixteenth century — we have already mentioned Frans Floris and, not to be forgotten, Jan Sadeleer! Jacques de Gheyn, together with Gillis van Coninxloo, is surely one of the most interesting and attractive artists of the generation which ushered in the seventeenth century and fecundated the North.

This Antwerp artist brought new life into the old relationship with Haarlem (as a pupil of Goltzius), and also into the Haarlem-Amsterdam axis. At the same time he was one of the most important links in the extension towards Leiden and The Hague with what was going on in the above places. He stayed frequently in all these places, and introduced vital innovations leading to interesting developments. We encounter him here not only as supplier and executor of an Antwerp motive, but in what follows as perhaps the chief founder of new motives and forms which were to develop into a means of expression typical of the northern Low

Countries and with which 'music in paintings of the Low Countries' is most closely connected.

De Gheyn probably settled in Leiden in 1593, and married Eva Stalpaert van der Wiele in the spring of 1595. Although De Gheyn was on a good footing with the religious reformation and gradually became increasingly accepted in the reformed circles of the Court and the aristocracy — his wife, in spite of being the great-niece of the priest and poet Johannes Baptista Stalpaert van der Wiele, came from the reformed side of the family — he still published whilst in Leiden the print of Saint Cecilia, thus demonstrating his Antwerp origins, not forgetting Sadeleer's picture motets.

What could have made him conceive this print? Rather than seeking the cause in the fact that he might have been able to sell it in Antwerp, or in the possibility that he had already made the painting or drawing of it at an earlier period (his Leiden pupil, Zacharias Dolendo[29]) made the engraving), we feel that it was made for a special occasion. The six-part motet was composed by the Leiden organist, Cornelis Schuyt, who was however a Roman Catholic, marrying in 1593 Cecilia van Uytgeest Pietersdochter. It is possible that Schuyt was acquainted with this type of print and that he commissioned De Gheyn to make a print, or perhaps first a painting of the patron saint of the organists and of his bride as a wedding present to the latter, which means the print must date from 1593. We might remind the reader that in Dutch aristocratic circles, especially in Amsterdam, Haarlem, The Hague and Leiden, the religious contradictions were not so important, and social intercourse among the various parties went on practically undisturbed. The text of the motet is remarkable: it concerns a humanistic correction of

Jacob de Gheyn and Zacharias Dolendo, S. Cecilia, engraving (Ex. Den Haag, Gemeente Museum)

Jacob de Gheyn and Zacharias Dolendo, St. Cecilia, detail.

the second part of the ancient liturgical antiphon, 'Cantantibus organis, Caecilia Domino decantabat, dicens (this part of the text is depicted by the organ-playing Cecilia): Fiat cor meum immaculatum ut non confundar'. This text, which also occurs in Domenichino's Saint Cecilia at the Louvre, is changed in this work to 'Domine fiant anima mea et corpus meum immaculata, ne confundar'. This correction in the classic Latin style is by no means a happy one. Is the corrector identical with the humanistic, Catholic poet of the hexameters underneath the print, who signed his monogram, C.D.? In any case, this picture motet, which shows on the right-hand side a vase containing the lily of purity, with above it a view of a room in which another shining heavenly messenger is crowning the martyr Cecilia and her chaste husband Valerianus, belongs completely to the same counter-reformational baroque atmosphere as do the other Antwerp picture motets.

Incidentally, the way in which this Saint Cecilia organist lived on with the Roman Catholic organists of Leyden after Cornelis Schuyt is shown by a small painting which was auctioned in London in 1937, signed A(ry) D(e) Voys: 'Saint Cecilia at the organ appears to Saint Luke'. Ary de Voys, a member of a family of renowned organists — he was one of the sons of Alewijn Pietersz de Voys — was born at Utrecht in ca. 1631 while his father was still the Cathedral organist there. Together with his oldest brother Pieter he learnt to play the organ from his father, who in the meantime, since 1635, had become the famous organist of St. Peter's Church at Leyden. Pieter was destined to succeed his father, and Ary, who also had a talent for painting, was apprenticed to the Utrecht master-painter Nicolaes Knüpfer — who was also

Jan Steen's teacher, and whom we shall discuss later on. The De Voys family apparently resumed earlier connections with Knüpfer, having a love of music and art in common. After further studies with Abraham van den Tempel, who had come to Leyden in 1648 — we shall see more of this master later on too — Ary entered Leyden's Saint Luke Guild in 1653. The picture, which, like the painter, links Saint Cecilia — musica, ars organi — and Saint Luke — pictura, and about which Houbraken tells us that the son dedicated it before his marriage to the father in gratitude, must have been painted in or close to 1653.[30])

Picture motets — if we disregard the one late Roman descendant of 1617 — were not done for longer than about ten years. They belonged nonetheless to the most special and attractive phenomena in the history of art and music. They contained the late mediaeval idea that music was for praising God and also an image of Eternity. But whereas in the Middle Ages the far-off heavens occasionally opened a benevolent window at which Angels appeared so as to let the people hear a little of the eternal 'Gloria in Excelcis', the people living in expectation of going there when their life was ended, the people who lived on an earth that remained earth, people who remained people, — in the baroque counter-reformation the heavens open up completely and descend to earth, filling it with dazzling light and transforming it into heaven on earth where the eternal song of God's praise is not the future, but reality.

Just as other phenomena of this cultural phase, Flemish picture motets are an expression of early baroque, counter-reformational triumphalism. This is confirmed in the 1617 print which seemed to fit so perfectly in the style of Rome, the heart of this movement. This boundless optimism, in music too, perhaps drowned the ancient pessimism that continued in humanism and 'profane' music; this attitude kept however its hard grasp on everyday life. This is why we must now pay attention to the sixteenth-century secular music of the South Netherlands that was prevalent in humanistic circes.

II The Good and the Evil Music

In paintings of secular sixteenth-century South Netherlands music, notes are few and far between it. It is nonetheless important to take a look at this domain, not only because it completes the picture of the entire musical life of the South Netherlands, but — and this is very special in this essay — because it is an indispensable introduction to understanding the 'Music in paintings of the Northern Low Countries'.

We have already discussed Maerten van Heemskerck's engraving (1554) and the so-called 'concert in the egg' (ca. 1560) which is in the museum at Lille. Both works are in different ways at the threshold between the late Gothic period and modern times. Both of them reach back into that recent past. Van Heemskerck has drunken clerics sing a polyphonous motet; the anonymous Lille artist also criticises the deteriorated monks. Both painters adapted this motive to the new age. But van Heemskerck was far more modern: he intentionally fitted the motive to an entirely humanistic, late Renaissance vision. The other one merely wanted to continue Bosch, and yet still included, unintentionally, something of the contemporary mood. The pessimism with regard to the music of the late Gothic period and the pessimism of the humanists

reach out towards each other, but differ totally. Bosch's satyrical rejection is the direct expression of mediaeval Christianity's dissatisfaction, which had increased to the point of turbulence, making it ripe for the Reformation; it is an accusation against the entirely changed relationships in an upturned world. Heemskerck's humanism had its roots in the intellectual world of the pre-Christian classic Antiquity, assumed into the humanism of the Florentine Renaissance and greatly increased after 1500 in the intellectual circles of Northern Europe.

One might conclude from Van Heemskerck's interpretation of music that music was in bad odour with the humanists, but this is totally incorrect. Van Heemskerck's theme is just one of the many and complex views formulated by classical antiquity and extended by its spiritual heirs such as humanism. Music is essentially good, even supremely so: it is the sounding harmonious arrangement of the universe, the macrocosmos, and of human internal and external relationships and virtues, the microcosmos. Music is harmony, even the principle of all harmony throughout. This is why children should primarily be educated by means of music. But Plato especially pointed out the equally great dangers connected with music: he also pointed out some bad kinds of music which led to the opposite of order — to chaos and destruction.

In the ideas about music in South Netherlands humanism, many facets of this are expressed in the music itself, but also visually, very clearly, in painting. Painters such as Pourbus, Floris, the various members of the De Vos family including the Martijn to whom we have already referred, saw and painted the execution of music as the expression of harmony in the family; as the expression of happiness too, it being the fruit of good harmony in the family. And especially as the expression of the richest fruits of harmony — love. Although our southern neighbours, in their sunny optimism, plainly preferred this good and carefree side of life and music, they indicated as good humanists in numerous works, both veiledly and candidly, the evil practices and effects of music. It is in love and the joy of life that the greatest dangers lurk: licentiousness and forgetfulness of life's frailty. Love and joy of life are supremely good, but only if they are tempered with moderation and contemplation of life's end. This accounts for the everpresent moralising, usually by means of emblems or symbols, which everybody understood in those days. Moreover, the intellectual nature of humanism was agreeably stimulated and activated by the occasionally ambiguous play of emblems and disguises, of 'pleasant obscurity' ('aengenaeme duysterheyt') (Cats), of 'profound finding and secret understanding' ('diepzinnige vonden en heymelycke verstanden') with 'one or t'other instructive meaning' ('d'een of d'ander leersame beduidenis') (Van Hoogstraeten). A painting without such accessories ('bijwerck') was not only dull, but also neglected its duty of providing a moral lesson.

Some artists expressed the pessimistic side more strongly; others exposed the evil side of music whilst taking a perverse pleasure in it. In general, the South Netherlands painters tended more towards a bright view of life. However, several influences caused the moralising warning that was part of humanism to be expressed more clearly and formulated in a typically Netherlands way. These influences included mediaeval morality, the counter-reformation of the Jesuits in the Flemish towns, especially Antwerp, and the contact with the more melancholy North,

in our case especially with Haarlem. In this way it was possible for creative minds from the North, such as Dirck Volckertsz Coornhert, the versatile Amsterdam humanist whose chief radius of activity was however Haarlem, to inspire by word and art the creative minds from the South and arouse them to far-reaching activity. This is how Antwerp in its turn presented to the North a flood of new vital ideas, motives and forms by means of masters such as Gillis van Coninxloo, Martijn de Vos, Jacques de Gheyn and the Ruckers family, each in his own way. We shall see how this happened in the second part of this essay.

At this point I should like to mention a print concerning good and evil music which clearly demonstrates the relationship between Haarlem and Antwerp. It was published by Cock in Antwerp, and is by Philips Galle — born in Haarlem, one of the intellectual circle around Van Heemskerck and Coornhert, but permanently settled in Antwerp — after a design by Frans Floris with the following text:

'Vt quidem magnetes ferrum attrahunt,
at theamedes qui in Aethiopia nascitur
ferrum abigit, —
ita est musices genus:
est quod sedat affectus, est quod incitat.'

'As some magnetic stones attract iron,
but the theamede stone from Ethiopia
repels iron, —
So is the genus of music:
There is music that soothes the spirit,
and music that incites it.'

The print shows a group of wild-looking people playing chiefly wind instruments, and somewhat apart a woman sitting in a dignified and beautiful attitude calmly playing a cittern; one of the others looks at her with desperate longing[31]).

Special attention, however, is attracted by a print which Crispijn van de Passe the Elder (a pupil of Coornhert) engraved at Antverp around 1590 after a painting by Marijn de Vos of 'Terra', 'the Earth'. Of all the prints which came from this circle and which were spread and imitated throughout Europe — depicting music and its function, such as 'Auditus' from the series of the Five Senses — this picture from the series of the Four Elements, is perhaps the most clear and significant one for us.

The interpretation of Earth in this engraving is basically favourable and optimistic: the Earth is a gift from God, and her fruits may really be enjoyed. Earthly enjoyment can be seen to be chiefly expressed by means of music. Music leads to the most glorious enjoyment — love, the harmony between man and woman. At the left we see fruit being plucked, a reminder of Paradise, but this is where the warning starts: it is not allowed to eat from the tree in the middle of Paradise, the tree of knowledge of good and evil. Well, in the midst between the principal figures playing their instruments so peacefully we see this tree with its fruits: the tree separates the two people who are craving to be united. The harmony of song and instrument is threatened. At the right we see the pleasant life in the merry month of May, ('den lustelycken mey'), again with music round the maytree, also leading to love, marriage and fertility; a loving couple sits in the shelter of the evergreen hedge. But here too is a conspicuous warning: the evergreen box-tree, symbol of youth, which can be seen over the shoulder of the young female principal figure, is at a slant and about to fall over.

On the table are the richest ripe fruits and other delicacies, and musical instruments

And although these emblems of a good and harmonious love-life contain a warning in themselves (ripe fruits rot quickly, music can lead to the wrong kind of love), the chief warnings can also be seen on the table: the lizard, the carnations and especially the open musicbook. The lizard first: apparently fleeting and unobtrusively present, the animal is the symbol of the incapability of falling in love — it cannot become ardent, and often means that love can cool off; it often coalesces with another emblem the salamander which symbolises something which might equally well be meant here: the salamander extinguishes the evil fire and nurtures the good fire — of love of course in this case. In both meanings it is a warning here: 'remember that love can end and disillusion', and 'keep love's fire pure'.

And then the carnation: emblem of love and eternity, here reduced to extremely pitiful circumstances — plucked without the stalk and planted in sand. This, too, indicates the irrevocability of transiency.

And finally the open music-book: it suggests a French lute tablature, but also gives the title of the piece, 'Susanna'. Two things should be observed here. First, this title is not intended for the musicians but is intentionally turned the 'wrong' way round to us, the regarders of the picture, and this leads us to the second point, which is that it has something to say to us. It reminded people in 1590, and still reminds plenty of us today, of one of Lassus' most well-known songs: 'Susanne un Jour'. It is the lesson of the chaste Susanna, waylaid by two evil elders, the lesson of pure love, which should be taken to heart here. The words of the song were not only famous because of Lassus' composition, but had been since 1548, when they appeared in the Premier Livre de

TERRA.

Humorum guttas mater cùm Terra recepit ;
Foeta parit nitidas fruges, arbustaque læta ;
Et genus humanum, parit omnia secla ferarum:
Pabula dum præbet, quibus omnes corpora pascunt,

Martin de Vos figurant.

Et dulcem ducant vitam, sobolemque propagent:
Unde etiam maternum nomen adepta es ť.
Luxuriem ergo caue partis quæ parcere nescit,
Donorumque Dei, quæ Terra hæc gignit, abusim.

Crispin de Passe sculp: et exc.

Martijn de Vos - Chrispijn de Passe, Earth, engraving (Ex. Den Haag, Gemeente Museum)

Chansons Spirituelles composées (in this case: versified) par Guiliaume Guéroult, published at Lyons. The text was set to music not only in this book, but later too[32]), a great number of times. In our engraving the title is 'Susanna', which means it is a Dutch version. There were several Dutch translations, and the one meant here was most likely:

'Susanna haar baeiende in een fonteyn
verborgen daer lagen twee ouders vilein,
al achter die hagen des boomgaerts stille.
Toen sij aensagen heur lichaam rein,
sij cregen behagen daar in certein,
en meinden te volbringen hunne valschen
wille.
Daerom sprack sy: 'Heer staet my by''[33])
Etc.

(Roughly: Whilst Susanna was bathing, two evil elders hid behind the orchard hedge. The sight of her pure body filled them with pleasure and they decided to fulfil their foul intent. She thereupon said, 'God be with me'.)

In one of the editions of the Livre Septième of around 1647 at Amsterdam, there was a setting by Dirk Janszoon Sweelinck of a text, the beginning of which is significant for our purposes: 'Susanna een reys verzocht tot wulpsche minne'[34]) (in effect: Susanna tempted to lascivity.) The engraving shows a tenor clef, appropriate to the lute, and the notes probably are based on a particular composition which De Vos and Van de Passe knew. We shall not go into the more detailed symbolism of the print

(such as the fruit and the carnations) here. The moral of the entire work, in which music plays a principal part, is classified again in the hexameters under the print: 'the earth nourishes and multiplies man and beast, thus granting them a pleasant life, and is as a mother, but beware then (ergo!) of voluptuousness which knows no mercy, and of abusing the name of God who created this earth'.

The 'Terra' representation by Martijn de Vos and Crispijn van de Passe demonstrates how musical compositions could be introduced to the emblematic world and function of visual art and also that even the music itself, being performed fulfilled an important moral function and expected understanding for this.

The prevailing optimism — complete in the church-life of the counter-reformation, relative in social life — had a natural breeding-ground in the South Netherlander's cheerful nature. He tended more towards a hedonistic attitude towards life and music, and this is hinted at everywhere. However, the stricter side of humanistic moralising had a stronger hold on the sympathisers with Calvinism, many of whom went to the North Netherlands. 'Susanne un Jour' was from a Calvinist source![35]) The counter-reformation, however, which had already taken its place in a baroque heaven, allowed the glad tidings of the eternal hereafter to overflow into everyday life: and so music in the South Netherlands was still mainly a symbol of eternal joy.

The North
Netherlands 1600-1690

The Preparation:
The Period of 'The Praise of God' and the
Change ca. 1500-ca. 1600

The earliest known painted musical compositions of the North Netherlands were
by Jacob Cornelisz van Oostsanen (ca.
1465/70-1533), citizen of Amsterdam, 'Jacobus Amstelredamus'. This late Gothic
mannerist, originally connected with the
Geertgen-tot-Sint-Jans area, but around
1510 becoming the first important North Netherlands artist to take Romanism in the
'language' of early Italian Mannerism into
his work and to go his own way, kept entirely to the Netherlands Gothic tradition
as far as music was concerned. He interpreted the intimate song of praise of the
angels with the intimate glad tidings
to the people at Christmas: it is still 'Gloria in Excelsis'. One of the most delightful
paintings with music is in the Capodimonte
Museum at Naples:[36]) the memorial tablet
of 1512 for Dirck Boschuysen, who died in
that year as the Lord of Offem, pictured
with his sister Geertruyt Boschuysen and
her husband, Jacob Pijnssen, fiscal advo-
cate at the Court of Holland, and now having become the lady and lord of Offem.
Dirck and Geertruyt are commended by
their patron saints. In the centre and foreground cherubim perform 'Gloria in excelsis...'. Four of them play the four parts on
wind instruments — two flutes (or shawms),
a trumpet and a sackbut, a fifth probably
doubling the tenor part on the psaltery. The
smallest cherub sits in the middle and holds
the book open. The Angel's Song of Praise
and the notes can be seen plainly: 'Gloria
in Excelsis Deo et in Terra Pax Hominibus
Bonae Voluntatis'. However, this polyphonic
composition, all four parts of which are
painted in accurate detail, is not, as is

generally the case, the beginning of the
Gloria from the Mass; but the *cantus fir-
mus,* and with it the entire composition,
is the antiphon with the same text of the
Benedictus from the Christmas Lauds, from
the Dawn office thus, which is of significance, as will be seen later. We do not
yet know who the composer was or where
the music comes from. Perhaps the painting can show us. It was intended as an
altar-piece, probably for the family chapel
of the lords of Offem in the church at
Noordwijk, which is near where the manor
was situated. Above the Holy Family is a
vista of the sea — the North Sea at Noordwijk? — with a citadel in the distance and
a harbour town on the left with bastions,
and also buildings probably imagined by
the painter — an allusion to The Hague-
Scheveningen or Noordwijk-Offem? — with
a landscape from the imaginary surroundings of Bethlehem transported here. It is
indeed the hour of daybreak: dawn can be
seen breaking from the left over the sea;
the hour of the 'shepherds' mass' and of
Praise: we can see shepherds already
hastened to the crib and still in the fields;
the last angel, announcing the Glad Tidings,
is already flying away. And the shepherds
join the Angels' concert, continuing the
'Gloria' at the crib where God is lying as
Mary's Child. Is this principal psalm from
the Office of Christmas Lauds and perhaps the entire book of psalms too, connected with the lords of Offem? Was it
part of the foundation? Was the music commissioned by the Boschuysen family, as
was frequently the case, another patron
being Pompejus Occo, who ordered music
for the 'Heilige Stede' chapel at Amsterdam a few years later? Must we seek the
composer in the Leyden-Haarlem circle or
in that of Amsterdam, or must we search

Jacob Cornelisz van Oostzanen, Birth of Christ,
details (Naples, Carpodimonte museum)

among the famous South Netherlanders or for one of them who worked in a town of Holland? In view of the fact that the music is an antiphon from the Laudes, it might be from a 'Zevengetijdenboek' (Seven Hours Book) like those of the Leyden Brotherhood of Hours singers which have been preserved.[37]) However, although the founders of the altar were able to find a painter in Holland, they will most likely have had to look for a composer, particularly in those early years of the sixteenth century, in the South Netherlands, where among the shining lights were Pierre de La Rue at Mechelen, Loyset Compère at Saint Quentin, Benedictus de Opitiis at Antwerp. One final remark: the painting links on to the 'Ave Maria, Mater Regis Angelorum' tradition discussed earlier.

As well as this great monument on a small document, there is also the tryptich by the same painter, made in 1515 for the Amsterdam humanist and banker, Pompeius Occo, and now in the Antwerp museum[38]). It is a devotional piece with Madonna and Child, again with angels playing music, including a group hovering high up in the sky singing from a sheet of music[38a]. This little domestic altar must moreover be regarded as the 'miniature pendant' to larger altars, in the style of the Van Boschuysen memorial tryptich, which Van Oostsanen was commissioned to paint in the same years by that same Pompejus Occo to decorate the chapel of the 'Heilige Stede' in honour of the Sacrament of the Miracle at Amsterdam. Occo vied in his many-sided benificence to the fine chapel with counts (of Holland), dukes (of Burgundy and Habsburg) and emperors (Maximilian and Charles V). He was also responsible for the sounding 'Gloria' and 'Laus Deo': first of all the famous organ dating from the same period and the case of which is in the St. Nicolaas Church at Jutfaas near Utrecht was in my opinion one of his donations; he also ordained at special Masses and Motets for worship in the chapel, and he sought the composers and writers of the precious psalmbooks in the South Netherlands too. Jacob Cornelisz van Oostsanen and Pompejus Occo were the keystones of the arch in Praise of God in the mediaeval South and North Netherlands.

But as early as the decade in which both died — in 1533 and 1537 respectively — the situation in Amsterdam had changed completely, and an end had come to the beautiful calm of devoted idealism. Not the increase of humanism, but the religious agitations made Amsterdam into a city of ferment. People inclined to reform were mainly to be found in aristocratic circles, among the rhetoricians and the squads of civil soldiery ('schuttersrotten'). The town was governed alternately by Lutherans and by Roman Catholics supported by 'Spanish' Brussels.

The next painting of interest to us dates from the thick of this period: Cornelis Anthonisz' 'Marksmen' portrait of 'squad D' called 'The Braspenningmaaltijd', of 1533. Among these members of the citizen soldiery, who often also belonged to the guild of rhetoric, were the most revolutionary elements. The most rebellious rhetoric plays (in the religious sense) were performed on the premises of the citizen soldiery. One of the most aggravating demonstrations against the old devoutness was the changed marksman's 'uniform': they assumed the brownish-red habit in mockery of the Franciscan monks, who occupied the most important position among the clergy in the centre of the old city. The marksmen of the Saint George company ('de schutters van Rot D

Jacob Cornelisz van Oostsanen, Mary, Mother of
the King of Angels, triptych (Antwerp, Royal Museum)

Jacob Cornelisz van Oostsanen, Mary, Mother of
the King of Angels, shutters

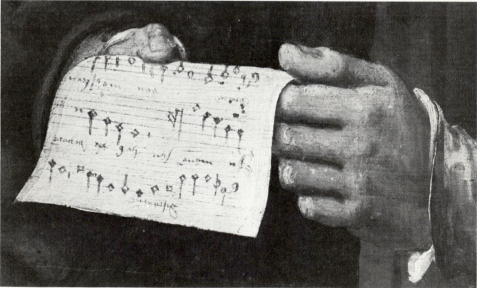

Cornelis Anthonisz, 'De Braspenningmaaltijd' (Amsterdam, Rijksmuseum: property of the City of Amsterdam)

der Sint Jorisdoelen') wore clothing, which was partly brown (from the Franciscan habit), partly blue[39]). In the picture they are sitting together round a table. One of them has a sheet of music in his hand on which the 'Discantus', (the discant part) of the song 'In mijnen sin heb ick vercoren een meijsken...' ('In my thoughts I have chosen a maiden...') can be seen, whilst his neighbour is taking a recorder out of its case, surely with the intention of doing the same with the other two recorders. The song had been extremely popular since the second half of the fifteenth century, and set polyphonically in songs and masses by many and great Netherlands, French and German masters: Busnois (died 1496), Alexander Agricola (died 1506), Hendrik Isaac (died 1517), Josquin des Pres (died 1521?) and others, Heinrich Finck (died 1527), Mathias Greiter (died 1550), etc.[40]). The Discantus — or 'Bovenzang', as it was called sometime later in the Northern Netherlands — painted here is the actual melody.

In   mij - nen   sijn   heb   ick   ver - co -   - ren.

Does the choice of this song, which is meant to confirm the harmony of the gathering, as does the banquet, have a special significance? It can hardly be otherwise in view of the very precise reproduction and the humanistic spirit of the time. The banquet was probably held on the festival of the patron saint of marksmen, Saint George, April 23rd. He can be seen in the top right-hand corner of the panel: a knight on horseback pictured in a stained-glass window. But this only is half the St. George story. The other half is that he rescued a beautiful young princess from a dragon. A landscape in which this is supposed to have taken place can be seen through the window, but there is no sign of the maiden, and it is not certain whether the painter really intended to illustrate the dragon's home-ground. For although this landscape is not reminiscent of Amsterdam, it was the fashion to paint such views through windows, especially in portraits of Amsterdam citizens, at that time. Perhaps the maiden for whom St. George is searching is the one referred to in the song, so that the marksmen could feel the association with their patron saint. And in the second place we see the merriment of the gathering expressed here. But perhaps there is a third 'meaning'. If we consider the brown habits, might we not see a gibe at the poor Franciscans, which would mean that this painting represents the same type as those from the circle around Hieronymous Bosch (the monks making music in the egg), where the lovesong indicates the depths to which the monks have sunk, or simply a gibe at their celibacy? The interpretation of this well-known painting is still a crux for art-historians today: the language of the gestures, the eating implements and the food is not yet understood; if it should ever be understood, perhaps the role of the music will become clearer.

II Towards the 'Vanitas' ca. 1600-1622

Up to this time, the appearance of music in paintings did not differ so very much from that in the South Netherlands. The dramatic development in religion and politics in the North, however — the separation from Rome and Spain by the reform, and the breakthrough of national awareness — caused a great crisis from which an entirely different climate emerged as a result of the cooler, more sober and solid nature of the people, which gave music a different type of expression, content and function.

The Southern dualism of musical ideas of the counter-reformation and Flemish humanism led to an airy compromise, because the two parties came together in agreeing optimism.

The Northern dualism of reformed Christianity and the humanism that was directed towards the Antique period reached a compromise that was as remarkable as it was difficult: an apparently impossible compromise. On the one hand there was the strictly orthodox Calvinism which only permitted church singing to honour God — the psalms, and which usually rejected and mistrusted secular music. On the other hand there was this secular music, which we might divide into three groups here: the performance of music in aristocratic circles, which fitted in with the humanistic ideas, — the music among the common citizenry with its North Netherlands realism, wordliness and frequent coarseness, — and the public performances of music arranged by the magistrate, bells and organ performances in churches of 'scandalous', 'voluptuous' and 'frivolous' songs by organists who had remained popish, being an abomination to the Calvinists. In the sixteenth century, this last-mentioned branch of secular music, the public performance, had most in common with the citizenry: it was a faithful reflection of it. In the seventeenth century, however, it became increasingly confined to the first two circles: of Calvinist and humanistic ideas.

As far as Calvinist music is concerned, it corresponded to Calvinistic 'iconoclasm' — the spirit which gave rise to 'image-breaking' — it was extremely sober and thus dismissed the visual arts; it was 'an-iconic' and thus the complete opposite of church music in the South Netherlands, that of the baroque counter-reformation, the fertile ground from which the 'Picture motet sprung. For our purposes, then, Calvinistic music itself does not supply us with any material. On the other hand, the attitude, or rather attitudes of the aristocratic humanists are extremely interesting: they are at the centre of the development of musical ideas in the North Netherlands and their representation in paintings. In spite of its great independence, this humanism had a distinctly two-faced attitude to the two extremes: Calvinism and popular culture.

Certain intellectual circles or personalities showed themselves to be increasingly sensitive to the sobriety and the pessimistic view of life of the Calvinists; this was because of one particular characteristic of the North Netherlander: his own feeling for moderation, solidity, his critical mind, his cooler nature. Others, however — but often the same ones! — exhibited an unconcealed attachment to the earthy, full-bosomed cheer of the citizenry — this was also because of another North Netherlands characteristic: they themselves had only recently risen from those levels.

This double-faced attitude among leading intellectuals involved a duality in expressions of art, and thus also in the role and ideas of music, and in its reproduction in paintings. In both attitudes a main motive crystallises, making them clearly distinguishable from each other: the expression of *transiency* in the stricter direction, and the expression of *love* and eroticism in the other. The North Netherlanders — with the aid of the South Netherlanders! — were brilliantly resourceful in both respects, in creating and working out motives. This is especially noticeable in the stricter direction: where

Calvinism certainly did not inspire design, but on the contrary rejected it, this direction gave us one of the most precious phenomena in the art of painting — the 'Vanitas', in which the process of incorporating music is in itself a fascinating occurrence, as we shall see presently. The characteristic of humanism, however, remained plainly visible in both directions; we can always see the roots, as well in the 'transiency', in the 'Vanitas' on the one hand and in the many and exuberant facets of eroticism on the other: the Antique attitude and emblematism and their development in the humanism of the Renaissance. This tree, grown from those roots, could not however safeguard itself from distortions imposed on it by the new impulses. This is perhaps less striking in the case of the North Netherlands expression of transiency in painting and music, where the Platonic ideal is preserved in broad lines: music as harmony of the 'careful' family, aware of the vanity and transiency of this life. (Not a trace of an alternative: a reference to the other life, the hereafter).

Highly intriguing, however, is the application of love emblematism: whilst het moral finger rules everywhere in this all-enveloping domain of life, a sometimes concealed, but usually obvious sympathy is visible and even characteristically in favour of the dangers and backsliding against which both Platonism and humanism warned so frankly. The inn, the brothel, the dissipated banquet are painted with all the emblems and indications of the perils; the 'wrong' music sings their highest praise, and Jan Steen takes the greatest pleasure in it all. 'Love' itself is reduced in coarse popular realism to the very thing moralism was warning us about: outright love-making, impudence and merry dissipation; these popular practices are placed on a pedestal with obvious exaggeration. This results in a peculiar kind of tension, which gives the North Netherlands painting of the seventeenth century a special position. However, this manner of representation only applies to extreme cases — which, incidentally, occur frequently; besides these there are all grades of less concealed immorality up to purely classical morals and even a mingling with the idea of *transiency*, occurring in varying degrees. The music on the paintings can clarify more to us in these fields — of transiency, vanitas and death, and eroticism — than all other means perhaps.

Symbolical of the important part which music was to play together with painting and especially on the Amsterdam-Haarlem axis were the brothers Gerrit and Jan Pietersz Sweelinck at the beginning of the palmy days of the North Netherlands. One of them was a painter, the other a musician, and both came from Amsterdam to learn their art in Haarlem; both were very involved in each other's art and in the development to be outlined in the following.

III   The Vanitas

The idea of 'vanity' is by nature connected with the essence of the art of painting: it is the transference of reality to shine. And what is shine other than 'vanity'? And also: 'Pictura quid est nonnisi vanitas?' Indeed, painting reflects reality. As sheer reflection, it merely leads to 'vanity', 'emptiness' (vanitas, inane). It should therefore have a different aim — *to confront* people with the real nature of things, with the fact that they are transient and vain. The purport of painting has always been firstly a moral one, before the aesthetic. This idea had for centuries been imprinted so strongly on people's minds as the prime meaning of painting that Jan Luyken as late as 1711 in his 'Leerzaam Huisraad', wrote a poem about 'The Painting' which began with the nearly obligatory words: 'De Schilderij is maar een Schijn, Van Dingen die int wezen zijn'. ('Painting is but a Shine of Things real'.)

The 'vanity' of painting becomes all the more poignant when it comes to portraits: to 'immortalise' oneself on a painting is a threefold deception — a *transient* person, who wants to display himself in his *vanity*, by means of a *vain method*. The remedy was soon put into the paintings: the emblems of transiency such as the hour-glass and especially the skull. The feeling for this was not as strong anywhere in Europe as in the Netherlands. As the epitome of wisdom, especially in the case of philosophers, was regarded the realisation of the idea of 'memento mori'. It is conspicuous that of the numerous imitations of Dürer's famous picture of St. Hieronymous as philosopher the very ones which were executed in the Netherlands especially those from the circle around Quinten Massijs, ca. 1520, and Lucas van Leyden, added a skull. Throughout the sixteenth century the portrait with the 'instructive' skull prevailed in North-West Europe, and especially in the North Netherlands. The serious, objective nature of the people most likely played a part in this.

There, too, a further development took place, taking form in the isolation of the 'memento mori' motive, the motive of transiency and vanity, to become the selfsufficient subject of the picture. The resulting type of painting was referred to as 'a skull', or 'deathshead'. The decisive phase was around the turn of the century — 1600. Once this idea of transiency had been isolated and raised to the level of an independent subject for painting, other transiency-motives announced themselves: the bubble ('homo bulla' — 'man is a bubble'), and especially motives from the Old Testament: the flower, crops. One could regularly read in religious texts what the psalms, Job and Isaiah had to say about it. Job, for instance: 'Man ... is of few days ... cometh forth like a flower ... cut down ... fleeth ... as a shadow' (XIV, 1-2). And Isaiah: 'All flesh is grass, and all the goodliness thereof is as the flower of the field: the grass withereth, the flower fadeth ...' (XL, 6-8). Peter and James had repeated and emphasised these words in their epistles. A little later the transient art of sounds — music — was added to this idea as one of the chief emblems. The development clearly shows the combined action of the skull, the flower and music.

We shall explain in more detail how this came about.

Strikingly, it was again the South Netherlanders who set the tone and found grateful disciples in the promising North with whom they brought their creations to full maturity. Once more the lifeline of Antwerp-Haarlem-Amsterdam proved fertile. Of the important Flemish painters who moved northwards under the pressure or not of religious upsets — Gillis van Coninxloo, Jacques de Gheyn, Roeland Savery, Jan Brueghel's far-reaching influence also making itself strongly felt — we must regard for our purposes Jacques de Gheyn as the greatest inspiration.

We have already seen how he contributed the last stone to the counter-reformational 'arc de triomphe' in honour of God and the Saints, 'the picture motet' in Leiden around 1593. But he was also to set the 'first stone' of the humanistic construction of the other camp. His life shows him to have been religious, but also 'libertine', and by no means disinclined to regard the ideas of reform favourably. Briefly, he belonged to the world of the Dutch humanists, to which, incidentally, his forefathers had also belonged. Although most of his work was done in Leiden and The Hague, we must first pay attention to a period during which he was exposed to decisive influences, which he later put into practice: his apprenticeship with Goltzius at Haarlem and two sojourns at Amsterdam (ca. 1587-ca. 1593 and ca. 1600-ca. 1604). In order to put our main objective in a stronger light, we shall have to make important sidetracks in our survey of the paths which De Gheyn took — sidetracks which will however turn out to be direct connections in general.

We ought first to take a look at his teacher, Goltzius. The vanity allegories had always been deepened and expanded in Haarlem, especially by the great masters, Coornhert and Van Heemskerck. The tradition was continued after them, especially by Goltzius, Van Mander and Cornelis Cornelisz. We have already seen how far-reaching their influence was in Antwerp on Philips Galle and others.

I have chosen a clear example of this humanistic Antwerp-Haarlem sphere of ideas: one of Goltzius' engravings which is of eloquent significance for music. It is one of the two portraits of young Haarlem officers which Goltzius engraved in 1582. The sheet is almost entirely filled by a figure of tall stature: an ostentatiously clad, well-fed and vain young man in the flower of youth is standing in a bare landscape, holding flowers in his right hand. One of the flowers is like himself, full of youthful strength, the other is already losing its petals. The words 'sic transit gloria mundi' surround them. The landscape enhances this lesson still further[41]). At the right we see a pleasant landscape decked out with green foliage, in which a company of people are enjoying themselves in the open air. This idyllic piece of nature is limited by a river. On the other side — 'our' side — where the young man is standing and occupying a far greater areas, everything is bare. But pay attention to the little bridge which spans the river: we see that there is literally where man 'transit', crosses over. Still young and in his glory, this man had to cross over, and is already in the land of earthly death. Under the print is a poem with the following closing lines: '... Dus, o idele, boos van gronden, betert u, want als een bloem is smenschen leven gestelt.' ('... O vain one, rooted in evil, mend your ways, for man's life is as a flower'). This is reminiscent of the psalm text: '... in the morning it flourisheth and groweth up; in the evening it is cut down and withereth' (Psalm 90, 6). (The counterpart of this portrait preaches the lesson 'Today: (so) — tomorrow: nothing'.) And what does this engraving have to say about music? During the period when it was made, from around 1580, increasing numbers of harpsichords made by the Ruckers family of Antwerp were sent to the North, especially to Haarlem and Amsterdam. (The reader should be reminded that Jan Pietersz Sweelinck travelled to Antwerp in 1604 in order to buy a Hans Ruckers instrument for the city of Amsterdam[42])). It is striking that it was the Ruckers house in which the idea of vanity, connected with music, and the moral remedy for this, was given form in a number of humanistic Latin sayings, three of which were to appear on their instruments far into the seventeenth century: 'Soli Deo Gloria', 'Sic Transit Gloria Mundi' (the real and the vain glory side by side!) and 'Acta Virum Probant'[43]). We shall discuss the

ONGHELYK IS TLEVEN der menſchen beuonde.
Deene tracht nae ſolaes, Ryedom, onnuttelicke ſonden
Dander door Gods gratie) alle booſe luſten velt,
Synde daer door verſekert nae Goodts vermonden
Teeuwich leuen Dus o idele boos van gronden
Betert v, want als een blom is ſmenſchen leuen gheſtelt

Hendrick Goltzius, Young Haarlem officer, engraving 1582 (Ex. Amsterdam, Rijksmuseum)

first of these a little later, since it flourished enormously in the North Netherlands around the middle of the seventeenth century. These texts were surrounded with countless painted flowers and smaller crops: emblems of transiency, of 'transire'. Ruckers' lesson was plain: make music, but do it for the glory of God, for otherwise music is vain and mischievous. We know that Goltzius, humanist and worthy successor to Coornhert, played several instruments, and also that he kept up close connections with Antwerp, and he must have been acquainted with these harpsichords and their motives.

De Gheyn, who according to Van Mander, practically abandoned his engraver's tool for the brush after 1591, first painted a *blompotken* ('pot of flowers') at Amsterdam, the first of a series. This was by no means new.

The vase of flowers in a niche — which is what is meant here — can be traced back over many contemporary and earlier Flemish masters to the fifteenth-century Flemish Primitives. But what De Gheyn did was not merely breathe new life into this genre of painting, not merely create a new centre for it — Amsterdam — but by assuming it into a special humanistic development of ideas of the period and surroundings, he made it into a new and more important emblematic point of departure which fitted into the general framework. These years keep on showing us how greatly De Gheyn — surely under the influence of Goltzius' circle — was occupied by the idea of transiency.

In 1599, the same year in which he also breathed new life into the 'bubble' motive using a print of Goltzius from 1594; the De Gheyn work is in the British Museum), and also an engraving 'Quis evadet'[44]), he made his will in his dwelling-place, Leiden, on August 31st, and the notary surely used De

Gheyn's own words when drawing up the will and writing that the couple is '... *de vijff Sinnen, na allen uytwendigen schijn* ten volle gebruyckende ... aenmerckende *des menschen leven als eene schaduwe verganckelicken* te wesen, ende *nyet sekerder* te hebben dan *de doot* ende *nyet onsekerder* dan de tijt.' ('to all appearances in full possession of the *Five Senses,* considering that *man's life is a fleeting shadow* and that nothing *is more certain* than *death* and nothing *more uncertain* than *time.'*[45])

In 1603 De Gheyn produced a new exemplary work, a 'deathshead': 'Democritus and Heraclitus'[46]), with the words 'humana vana' at the top. Abraham Bloemaert, the Utrecht painter who knew Amsterdam so well, must also have painted a 'deathshead' in the same year for a member of the Amsterdam humanistic circle.

Goltzius, as well as De Gheyn, were acquainted with an 'Allegory on Vanity, Death and Resurrection' by Martijn de Vos, at any rate with the engraving of it which Sadeleer made. It shows two putti, one of them asleep near a skull on ears of corn, the other playing with soap-bubbles among vases of flowers. Sadeleer's departure from Antwerp dates De Vos' painting before 1587. Among the spiritual inspirers of all such humanistic motives in the Counter-Reformational circles, the following must be particularly stressed: the famous philologist and philosopher, Justus Lipsius, who was the soul of a neo-Stoicism, professor at Leyden(!) from 1578 to 1591 and at Louvain from 1592 to 1606; members of the Van Est family of Gorinchem who also belonged to the Louvain circle and had personal connections with our artists — Franco Estius composed many a Latin rhymed caption for Goltzius prints.[47]) The vase of flowers in the niche, the 'homo

Harpsichord by Andries Ruckers, Antwerpen 1618
(Berlin, Institut für Musikforschung)

bulla' and the 'skull' were popular in Amsterdam during the years 1600-1605, and formed a closed group of genres which owners of picture galleries tried to acquire examples of, resulting in several commissions for painters who were well-known or, to us, less well-known. Among the famous art-collectors in question at Amsterdam were the notary, Jacques Razet[48]) and the book-dealer, Reynier Anthonissen. Painters known to have worked in Amsterdam at that time and to have carried out such commissions were the Flemish Roeland Savery (a flowerpiece from 1601 has been preserved[49])), David Bailly from Leiden (of South Netherlandish origin) and Cornelis van der Voort (born at Antwerp), who was Bailly's teacher, and in whose inventory of 1614 a 'flower vase' and a 'Democritus and Heraclitus' are mentioned[50]). Roeland Savery is of especial significance, because he was employed at the Habsburg Court a little later, and his journeys there and back caused this Amsterdam-Antwerp idea to spread, as we shall see. Bailly and Van der Voort were also important for Holland in continuing and imparting it in their circle.

It must have been during these years that music in Amsterdam humanistic circles was added to the idea of transiency which was then at its zenith. And there is absolutely no doubt that this was brought about by a contribution from no less a person than Jan Pietersz Sweelinck. This musician, whose greatness dominated all Europe at the time, had particularly close connections with the world of art and painters. It was the painter Van Mander who called him the 'Orpheus of Amsterdam', and his painter brother Gerrit 'the best pupil' of Cornelis van Haarlem[51]). At Van Mander's deathbed in Amsterdam on September 2nd 1606 were Jacques de Gheyn, who made a drawing of him[52]), and the Jacques Razet referred to above, who spoke the prayers for the dying: 'In manus tuas Domine commendo animam meam'. Van Mander had dedicated part of the 'Schilder-Boeck' to Razet. Razet's gallery included works by Gerrit Pietersz (Sweelinck)[53]). This gallery was such an important focal point that it was visited daily by art-lovers. We can surely assume Jan Pietersz Sweelinck to have been a well-known visitor there.

An especially valuable musical testimony of Sweelinck himself, from 1608, speaks for this. In the second half of May of that year he travelled to Harderwijk to examine, together with other prominent organists, the organ which Albert Kiespenninck had restored. On May 24th[54]) he was the guest of Ernest Brinck, a member of a patrician family of Harderwijk and later mayor of the town. Brinck (1582/1649) kept an Album Amicorum, famous because of the contributions from eminent personages, scholars, politicians, artists from all over Europe — including Galilei — and also proof that he was well acquainted with the circle of people discussed here. The album contains a 'bon voyage' wish from Roemer Visscher in 1612, emblematic drawings by Goltzius, Jacob Matham and Jacques de Gheyn, and the canon by Jan Pietersz Sweelinck on fol. 231r of Album I reproduced here: 'Vanitas vanitatum et omnia vanitas'.

Why did Sweelinck choose to this canon, which was plainly intended to be sung immediately by the cultured company at the house of the young man of letters, Brinck, on May 24th 1608 — the pessimistic opening words of Ecclesiastes? Such a text does not appear to us to be directly appropriate to a festive dinner, interspersed with witty intermezzi! We have to recall here the mentality of the sixteenth century and the first half of the seventeenth: the perpetual need to create a moral corrective as a counterweight to worldly pleasures. The portrait of the deathshead is more than a parallel here: the canon *is* itself such a portrait: Sweelinck's image, representation and signature, Sweelinck, the famous musician who drew and immortalised himself — recognising with his entire period: all this is mere vanity. The text is the deathshead belonging to his musical portrait.

From Ernest Brinck's Album Amicorum (Private Collection)

But it is striking that he expresses this 'deathshead' by means of an idea which is a precise echo of what was of focal interest to Amsterdam circles of painters and humanists.

However, there is more to come.

For the first time, the context of the increasing Amsterdam motive of vanity contained the chief catchword: Vanitas, the word which a little later was to denote the genre of painting of the morality of transiency.

This remarkable and striking choice is not just an echo and a side-effect of the painted motive, but supplements it, enlarges it and becomes united with it.

'Vanitas' refers for the first time to music: music, that excellent art, is itself with its fading sounds the prototype of transiency and vanity. Sweelinck makes music indict itself as 'vanity of vanities'.

The years that followed were to prove that this musical morality of vanity was connected to some purpose with the illustrative art of painting: Sweelinck's 'Vanitas' canon was the start of a development which finished with paintings of music as an emblem of vanity, with a Vanitas consisting of painted music and instruments produced in 1622 in Haarlem by Jan Albert Ban and Dirk Matham. Shortly after the maturing process of the 'skull' and 'flower' vanitas had been completed, the musical vanitas became a part of it. Before this we shall consider some of the intermediate stages; fortunately there are important documents to help us. In 1615, the year exactly between 1608 and 1622, a lute-book appeared in Nurenberg; we shall first discuss its title-piece, since it connects up better with the Vanitas canon, although important documents had appeared in Amsterdam as early as 1614.

Georg Leopold Fuhrmann, engraver and publisher of Nurenberg, published in 1615 the

Georg Leopold Fuhrmann, Testudo Gallo-Germanica,
Nurenberg 1615. Titlepiece (Ex. München, Staatl.
Bibl.)

'Testudo Gallo-Germanica', a lute-book in German tablature, but also, for the first time in German countries, in the French-Netherlands tablature; it contained compositions from practically all European countries, including Poland. Our chief concern, however, is for the title-piece. In the middle is a lute, with an open tablature book a little to the left, more in the foreground. The letters at the frets and also the music are in the French-Netherlands tablature; only above the last stave are there tablature signs in the German and Italian tradition. The emblematic-moralistic language starts however around the open lute-book: we see four plucked flowers, a sprig of lily-of-the-valley, two of sweet-briar and another delicate little plant: the 'faenum', the light field-crop doomed to 'exsiccare', to wither away. The transiency and vanity of music!

In the regions of South Germany too? The flower as a symbol of transiency was common in Europe, but not so the musical connection. We shall see that 'faenum vanum'[55]) in connection with music indicates the Netherlands. Proof of this is supplied by the upper right-hand side of the drawing. A heavenly hand reaches out from the clouds to point emphatically, like to some key, to the deeper, real meaning, to a stone niche containing a vase around which the words of Ecclesiastes, 'vanitas vanitatum et omnia vanitas', can be read. This is directly reminiscent of the Amsterdam niche with the vase of flowers as a symbol of transiency. But if we take a closer look we see that the vase's body is a skull, and its handles are two serpentlike worms which have wound their way up from the eye-sockets. The flowers are represented by plumes of smoke. The skull which the worms are eating away was an advanced stage of the 'deathshead' motive; this had already been done in the De Gheyn circle, but was more at home in the fiercer South Germany of the early Baroque period, with the atmosphere of mystic occultism that emanated from the court of Rudolf II. What was new was the combination with the flower-vase motive of transiency. The volatile, evil-smelling smoke which rises from the skull-vase instead of flowers is the interpretation of many a text from the Old and New Testaments such as 'For my days are consumed like smoke, and my bones are burned as an hearth. My heart is smitten, and withered like gras' (Psalm 102, 3-4), and 'For what is your life? It is even a vapour, that appeareth for a little time, and then vanisheth away' (James IV, 14).

The paths in the Netherlands and in Amsterdam in particular which led up to this representation are numerous.

It is not by chance that the composition was made in Nurenberg. In 1604, Roeland Savery (from Amsterdam) and Jan Brueghel (from Antwerp, but he most likely went together with Savery) left for the court of the Habsburgs, where Rudolf II was the life and soul of the exciting court culture. Their journey took them by way of Frankfurt and (1607) Nurenberg. We have already seen that Frankfurt was a centre for artists who had left the Netherlands, especially engravers and publishers. There was not a travelling Netherlands artist who would have missed a visit to 'Netherlands' Frankfurt. We have already mentioned that Jan Sadeleer was there on his way to Munich in 1587; Gillis van Coninxloo was there from 1585 to 1595 before his Amsterdam period. Savery and Brueghel were followed a year later by David Bailly, who also set out from Amsterdam, taking pretty much the same route via Frankfurt, Nurenberg and Augsburg towards the South — his journey ended in Rome

These masters also took the vase-in-the-niche idea to Central Europe.

They were fortunate that this very motive had already been prepared in central Germany, to wit, in Frankfurt, by their compatriots. After all, it was in accordance with Adriaan Collaert's example of publishing flower-vase engravings in Antwerp in 1596 that the De Bry family, South Netherlanders who had settled in Frankfurt, started engraving prints after paintings by Jacobus Kempeneer, six of which were published in 1604 in Frankfurt by another Fleming, Jan Busschemaecker. Sadeleer knew these prints, and published them in Venice before 1600. They are flower-vases with a simple indication in slanting, rising lines of a niche which is not worked out in any detail, with a moral Latin text underneath. The text of the first of these 'Polyproton de Flore' is: 'Flos speculum vitae modo vernat et interit aura . . .' ('The flower is the mirror of life; it has scarcely bloomed before the wind interferes . . .'). The manner of the representation here is however very dependent on Savery's style, with the bordered niche, and on Brueghel's as far as the shape of the vase is concerned; this is reminiscent, in spite of the skull, of the Italian late-manneristic elaborate vases of Fontana or the Saracchis which were popular (and imitated) at Antwerp, especially in the southern German states and at the courts of Prague and Munich. An illustration of how long the taste survived in Frankfurt for the Dutch Vanitas of the De Gheyn school with the vase in the niche, the skull, the homo bulla, music etc., is a painting by Jacob Marellus, — who was formed in Frankfurt and Utrecht, where he had known Savery for many years —, executed in Frankfurt in 1637[56]). There are more indications of the Netherlands and Amsterdam.

Fuhrmann's lute tablature of 1615 is not separate from the renowned editions of lute-music which saw the light of day in the Netherlands between 1600 and 1615: the lute-books of Joachim van den Hove of Antwerp, 'Florida' appeared at Leiden in 1601, 'Delitiae Musarum' at Utrecht in 1612 and the famous 'Secretum Musarum' of Nicolas Vallet at Amsterdam at the beginning of 1615. All these editions, together with those of Jean-Baptiste Bésard, who went to Cologne in 1603, later to settle in Augsburg, were the reason for Fuhrmann's introduction of the French-Netherlands tablature in the German region in 1615.

In this connection it is also interesting to note how well-known Sweelinck's works were at Frankfurt already during his lifetime: Georg(ius) Draud(ius) compiled catalogues of music at the behest of Frankfurt booksellers; in the section called 'Bibliotheca Exotica' from the first series of catalogues which appeared at Frankfurt in 1610-1611, recent editions are listed of works by Jan Pietersz Sweelinck from the year of his Vanitas canon: 'Livre Septième des Chansons Vulgaires, 1608 apud Corn. Nicolai', and 'Nieu Chyterboeck, tot Amsterdam bij Janson 1608 in 4'[57]).

The engraving is signed on the left, above the niche: 'Hauer fecit'. Johann (Hans) Hauer (1586-1660), painter, engraver, etcher, gold- and silversmith, art dealer and writer from Nurenberg, had many interests. He is known to have made elaborate goblets ('Prunkbokalen'). His characteristics are demonstrated in the title-piece, which his fellow-townsman and colleague Fuhrmann and invited him to make[58]).

To summarize, we can state that the clear connection between music and the flower in the Vanitas domain, the application of the classical Ecclesiastic 'Vanitas Vanita-

Johannes Torrentius, Still-life, 1614 (Amsterdam. Rijksmuseum)

Torrentius, Still-life, detail

Roemer Visscher, Sinnepoppen 1614, Titlepage

tum', the knowledge in the German region of Sweelinck's music, all speak for the opinion that in Hauer's title-piece we have a consequence of the development in Amsterdam and of the part played in it by Sweelinck. In any case, a total identification of music and the musical instrument as Vanitas in the art of painting had not been achieved, but was already more than half-way there.

The completion of this process, which was achieved after 1620, depended on other developments too. The most important group of contributions, including a painting which is one of the most interesting examples of 'Music in Paintings of the Low Countries', dates from 1614, to which we have already referred.

It is an unusual painting, this still-life from 1614, by an extraordinary man, Johannes Torrentius of Amsterdam[59])!

It was discovered in 1913 as the lid of a currant barrel with the seal of Charles I of England on it. Parallel to this: its painter, who was imprisoned and tortured in Haarlem because of his 'immoral' paintings, to be released by Prince Frederik Hendrik and Charles I of England, whose court painter he became.

It is the painting with the first North Netherlands musical composition on it, with remarkable and mysterious backgrounds.

It is a unique work, if only for its artistic qualities: with a chiaroscuro and monochromy which was not to be seen until Rembrandt, twenty years later, with a 'Feinmalerei'-technique which was to be realised by the Leiden School of Dou and his contemporaries only thirty years later, of perfect balance, elaborateness and expression of the material which were unparalleled, with a bold and general vision which broke through the bounds of the seventeenth century and even of any particular period throughout.

The connection with Roemer Visscher's 'Elck wat wils' print, which appeared in his 'Sinne-poppen' of that same year, 1614, is familiar and has been discussed several times, but by no means completely. The music which Torrentius added to the components has never been adequately discussed, however it can be of special help in illuminating the still-life. This is even more unsatisfactory because of the fact that people always recoiled from taking seriously the problem tof Rosicrucianism, a problem chiefly posed by Rehorst[60]). Neither the extent of this essay nor my competence allow me to go into this more closely here — but I cannot entirely avoid it.

Not only the strange cartoon ascribed to Adriaen Pietersz van de Venne[61]), and the letter from the Haarlem magistrate to the Hove van Holland dated June 19th 1625[62]) but also the music in the still-life refer to Rosicrucianism, and to Roemer Visscher and Torrentius as chief members of it.

For both of them the representation of the wineglass ('roemer') between the decanter and the water jug must have been the core of their idea and the year 1614 must have been a decisive stage in the growth of this idea, most likely in connection with the Brotherhood. (For the Rosicrucians the year 1614 must have been the beginning of a new era.) The 'Sinnepoppen', which appeared in 1614 bear the picture with the words 'Elck wat wils' as title-piece (the words might be said to mean today: 'as you like it'); this was already Roemer's motto as can be seen in Ernest Brinck's 'Album Amicorum', in which, as we have already seen, Roemer Visscher entered his contribution. Torrentius preserved his work as a treasure, rescuing it from all the imprisonment and torture so as to be able to present it person

ally to his saviour, Charles I, who had his seal placed on it.
Torrentius' additions and his alterations compared with Visscher's emblem serve to illuminate many aspects, but especially the music.

In the first place, Torrentius extended Visscher's horizontal composition by means of a vertical one, resulting in a ·cross, the music being its fundament. Torrentius changed two things in Visscher's emblem: the glass was made much bigger, and the decanter and water-jug were placed crosswise with regard to each other.

Let us start with the music.

Apart from the general praise of old and new writers, starting with Huygens[63]), there is only the following to be read about Torrentius' sheet of music: 'a verse with a far-fetched moral' (Schmidt-Degener in his guide to the Rijksmuseum[64])), 'Only in order to accentuate in a harmonious manner the shadow cast by the wine-glass was the top line of the music given the form of a garland ... There appears to be no musical connection between the notes and notation. The notes have merely a purely decorative character' (Rehorst)[65]).

The writer of the 'Zinne- en minnebeelden'[66]) is silent about this material which is so rich in emblematism and indispensable for comprehension.

In modern notation, the music would look like this:

ER   Wat bu - ten   maat be-staat   int   on-maats qaat ver - ghaat.

The extreme precision and clarity of Torrentius' musical notation ought to have warned interpreters that he painted the music in this way in full awareness.

Is it the closing fragment of an existing piece of music? The first note, which looks as if we had just interrupted the piece, might lead us to believe so. The nature of the text is found in works by the poets Spieghel and Roemer Visscher. But it might equally well be an independent composition, composed for this painting, perhaps, or perhaps not, as a result of the gatherings in the 'Roemershuys'. A closer analysis might serve to confirm this last assumption. The first semibreve, B, is conspicuous in having the letters ER underneath it, which, however, seems not to be part of the text. It is also strange that the piece ends on an F, thus appearing to be in that key, but the two B's contradict this: there are no flats or key-signature, which brings us back to the assumption that this is a closing fragment. A first step in the right direction is taken when we look at the arrangement of the 'bars', the duration relationships. The rest has a strange position between the words 'buten' ('out of') and 'maat' ('measure'), being proportionally discrepant there; there is a lack of balance between the two parts of the sentence which are meant to balance each other. If the rest were omitted, the 'equal measure' would be restored:

Wat bu - ten maat be - staat   int   on-maats qaat ver - ghaat.

But it is as plain as pikestaff that this is what Torrentius wanted; the rest has resulted in an expression of the text: the music is indeed 'buten maat' (out of time). This makes the second part 'onmaats' ('without time') because it has to continue out of step, as it were, and this leads to a 'bad result'!

But we still have the troublesome B. If only it were a B-flat, the music would sound very well. With disconcerting faithfulness, all writers have tried to correct Torrentius: they write 'quaat' or 'kwaad' (bad) where the painter has 'qaat'. But this word is under the B which ought to be a B-flat! This is clearly a double 'symbolic mistake': the note B causes a 'diabolus in musica', the worst mistake in music since the middle ages, and he *writes* the word meaning 'wrong' or 'bad' — 'wrongly': 'qaat!' We see in this a continuation of the late mediaeval musical symbolism, which, for example, used to express the word 'peccatum' (sin) by a musical sin.

Now, however, the significance of the first B becomes clear: it is like a second 'key', a warning: 'Take care, this is not a B-flat, but a B'. Not a musical note, then, but a symbolic one, a sign. But it does belong to the music in a different way: otherwise Torrentius would not have put a rest between this B and the G that begins the melody. This puts the B in a rhythmic relationship to the rest of the music. The result is a piece of five 'bars', five semibreves long. Each semibreve bar consists of three minims, a prolatio, with the possibility of imperfication for the semibreve, as also occurs in the first and third semibreve bar.

Five 'bars', then, of *three* beats. Irregular? The basic factor of Rosicrucianism is 53, the 5 and the 3, with their difference, 2. Our piece consists of 5 bars of 3 beats, and

Torrentius intentionally divided the 5 bars into 3+2 bars. I may definitely not speak with absolute certainty here, but it is food for thought and further investigation.

Much is still an enigma, such as the mysterious letters ER and the sign after them. Since the 'thirties'[67] it has been generally assumed that the letters with what is perhaps a cross after them are an abbreviation for Eques Rosae Crucis[68]). This is not impossible. But a closer look at the letter R shows that other letters are hidden in it: an L, a T, perhaps more? Is that really a cross after the ER? Or is it a concealed cross, in accordance with the methods frequently employed in occult circles? There is a cross at the end of the text.

More questions: we already noted the form of the cross in Torrentius' composition: did he intentionally place the decanter and jug crosswise? There was sense in the opposition by Roemer Visscher as explained by Dirck Pers in 1644 in his translation of Ripa's Iconologia[69]). There must also have been sense in Torrentius' alteration. And the big wineglass? The Dutch word for this is 'roemer', and the 'great' Roemer Visscher followed his own motto of 'something for everybody' and made a play on words with the wineglass and his own name. Does the enlarged 'Roemer' allude to the painter and also express something 'buten-maats' — 'extra-ordinary'? Filling it a third full with wine is certainly not 'onmaats' — 'without proportion'!

We are clearly faced with an emblem of Temperantia — temperance. The music also contributes to this interpretation: it is by definition an art of proportion, balance and harmony. The normal expression for vocal music used in the Netherlands of the seventeenth and eighteenth centuries was 'zang-maat' ('song-measure'). Music is placed here as the very fundament of measure, of moderation, and the text speaks about this phenomenon. It is obvious that we do not have to do with normal 'moderation' here. At present the 'buten-maatse' ('extraordinary') aspect cannot be cleared up any further, but it does provide more play for something which is beyond the normal proportions. And, in view of the secrecy that was normal in circles such as the Rosicrucians, might not the ER have something to do with this? Perhaps this ER displays the role of a key like the note B above it? Music is 'measure' ('maat' = ratio'): this 'ratio' is a fundamental and central idea of the still-life; Roemer Visscher speaks in his text to the emblem about the 'juyste even mate', which in Latin is: 'aequa ratio'. Might ER not be an abbreviation for the Latin for 'buten maat' — 'Extra Ratione'? And does the picture perhaps preach 'temperance' only *apparently*, and in *reality*, albeit in a conceiled fashion, refer to the extraordinarity of the man and Rosicrucian Johannes Torrentius and maybe his supporters?

'That which is extraordinary has an extraordinary bad fate' from the Rijksmuseum catalogue would then provide the correct meaning for the first half, but not in the second: Torrentius surely had a reason for using the two terms, 'buten maat' and 'onmaat'. The meaning could be: 'That which is extraordinary, beyond the normal measure, special, comes to an especially bad end if done immoderately with regard to the proportions'. This would also account for Torrentius' excesses in his life! During his trial he tried out the truth of this statement neither confessing nor bending: he 'bestond buten-maats' — he was 'extraordinarily' steadfast!

Torrentius' still-life is not only the first work with a definite North Netherlands musical composition, it is especially the first and one of the extremely scarce ones of its kind, because the notes of the music form an integrating, functional component of the painted composition and its actual meaning. The music here is not a 'Vanitas' as interpreted by Sweelinck, but 'Temperantia' and 'Ratio', with a comparable moral idea. The music is the basis of the 'temperance', which culminates at the top of the wineglass in the bridle which Torrentius adopted from Ripa's 'Temperanza' — its purpose was to 'rafrenare' — to curb. (Ripa's 'Iconologia' appeared in 1603!)

This work is not a 'Vanitas', but did play a part in the musical 'Vanitas' in 1622, as we shall see. We should note that it was in the same year, 1614, that the term 'vanitas' made its first appearance in literature as an indication of type, to wit in the descriptions of paintings compiled by the Amsterdam painter whom we have already mentioned in this connection, Cornelis van der Voort[70]). Torrentius was acquainted with the Sweelinck brothers, Jan the organist and Gerrit the painter, whom he surely met at Roemer Visscher's:

'wiens vloer betreden wordt, wiens dorpel
is gesleten,
Van schilders, kunstenaars, van sangers
en poëeten,'

'whose threshold was crossed by painters, artists, singers (here pars pro toto: musicians) and poets' (Vondel[71])).

The connection between the use of music in his painting and the role and influence of Sweelinck is still an obscure one, although very probable.

After Roemer Visscher's death, in 1620, and after his own divorce in 1621, Torrentius took up residence in Haarlem. Besides his

Dirck Matham, Vanitas, engraving, 1622. With the
poem and music by Johan Albert Ban (Ex. Haarlem,
Gemeente archief)

fellow-painters, including members of the Matham family, he had other good friends there, such as the merchant and art-lover, Isaac Massa, who got married at this very time (1622), — on which occasion he and his bride were painted by Frans Hals[72]), — and who lived most of time at his country-house near Lisse. Isaac Massa was also a friend of the Mathams.

Without going further into the thorny question of Torrentius' trials, it is important for us to establish the fact that his paintings, including the still-life with music (1614) were kept at Massa's house for a certain length of time. It cannot be proved that the Mathams saw them there in 1622, but it is certain that they had already known of their existence in these circles as early as 1622. This is proved by the work which we shall discuss next: the first musical 'Vanitas' (1622), an engraving by Dirk Matham, published by his father, Jacob[73]). Dirk was 16 at the time, so that part of the invention must be put down to the father, but also to the man whose part in the engraving is impressed twice on the copper: 'I. A. Bannius', who was the 24-years-old priest-musician, Jan Albert Ban.

In this detailed work, the first one to be called a 'Vanitas', by the extremely youthful step-grandson of Goltzius, who had died three years before, we see chiefly two planes which are connected with each other. Through a double window-frame we have a view of a room situated higher and further away in which a richly-clad company is banqueting, raising their glasses to the accompaniment of music played by a group of musicians sitting on a balcony. The 'vanity' of these worldly pleasures is literally emphasised and endorsed by the cartouche with a skull and the title, 'Vanitas', and it is also emblematically explained by everything in the foremost plane and also at the left. The wine-flask competes for importance with the musical objects: a five-stringed gamba, fastened at the top by a curtain-cord, a lute, a cittern, a flute and an open music-book. The wine-flask is the one from Visscher's Sinnepoppen and Torrentius' Temperantia. Here its emblematism has been changed to a certain extent: the water-jug has changed position — there is a tankard in the same style as the wine-flask on the left on top of the cupboard — there are two glasses, one full of wine, the other upturned and empty, in accordance with the commentary of another Haarlem painter, Pieter Claesz, on a Vanitas: 'The glass is empty. Time is up'.[74]) There is also an open chest with gold coins and jewels, and on the cupboard a cup with a cherub's head, indicating the fleetingness of life, death's cutting off a life which has scarcely begun. A third explanation, explicit now, is provided by the two-part song in the book, the text of which is continued under the engraving.

The poem, which states what the engraving depicts ('ut pictura poesis'), appears to be a kind of 'collage' of antique fragments, supplemented and completed by Ban. After all, the opening distichon, 'Omnia sunt ...ruunt', is taken from Ovid's Ex Ponto (IV,3, 35-36), which had already been used before Ban, for example by Philippe de Vitry around 1315 in his motet 'Quoniam secta latronum', included in the Roman de Fauvel.[75]) The end, too, 'caetera mortis erunt', is derived: It occurs on the title-leaf of Jacob Hoefnagel's 'Archetypa' (Frankfurt am Main 1592), likewise in connection with the emblematism of vanity — the 'homo bulla'. Ban's piece is a Baroque contribution to the humanistic ode-composition. The music is of an illustrative character which

'Everything that is of man hangs by a
thin thread,
And collapses by an unexpected fall.
What are: world, pleasure, vain delight?
Defection, pain, smoke and shadow:
nought.
Nor Gold nor ivory, food, drink, music
nor acclaim,
(Which quickly pass) together, cause
happiness.
Time and *eternity*, these two are our
true possession:
This man must consider — the rest is
for death.'

Omnia sunt hominum tenui pendentia filo,
Et subito casu quae valuere ruunt
Quid mundus, quid deliciae, quid vana
voluptas?
Faetor, tristitia, fumus et umbra, nihil.
Non aurum, nec ebur, cibus et potus,
organa, plausus
(Cum pereant celeri tempore) juncta beant.
Tempus et *aeternum* duo sunt quae nostra
vocantur,
Haec meditanda homini, caetera mortis
erunt.

he was later to describe as 'moving
the senses, soul-moving' ('zin- en ziel-
roerende zang'). Here we have a
double Vanitas: the first musical Vanitas in
painting, and a 'Vanitas' in music itself, the
interval gap and the broken-off series of
notes in the second line being especially
conspicuous.[76])

The engraving seems to have been inspired
by two other sources besides the Amsterdam 'Vanitas' development with Sweelinck's
first musical Vanitas, and Torrentius' musical Temperantia; one of these sources is of
Antwerp: Jan Brueghel's series of the 'Five
Senses' from 1617 and 1618 (more of which
later), the other source being Roman: Cara-

vaggio's paintings with music from the last decade before 1600.

As far as Antwerp is concerned: it is obvious that the Goltzius-Matham family and Philip Galle's sons, Theodor and Cornelis, continued the old Haarlem tradition of a lively contact with Antwerp; during these years, Jan Brueghel's fame had reached its zenith, and his work was much sought-after by engravers.

And as to Caravaggio: in the very years when he attracted attention with his first works in Rome, including 'Una Musica' (now at New York), the 'Lute-Player' (at Leningrad) and 'Rest on the flight to Egypt' (at Rome) — 1593-1600[77]), not only Jan Brueghel was in Rome, but also Jacob Matham and his friend Frans Badens for a fairly long time, as testified by his pupil, Adriaen van Nieulant.

Perhaps a sketch which Jacob Matham made after a Caravaggio was a model for his son Dirck. As for Ban, who was completing or had completed his studies for priesthood during these years, he might very well have travelled to Rome.

At any rate this is a remarkable youthful work, which took the transiency idea in music a decisive step further in the Dutch Vanity representations, thus completing and establishing the musical Vanitas.

Finally, we should like to attempt to bring the engraving a little closer to Torrentius' still-life. The use of music in the Rosicrucian work was striking, new, unique of its kind, and as we already stated, concealing and revealing, containing the heart of the idea. The work by Ban and Matham, so full of detail and so explicit, has also remained on its own. Is it a daring idea to think that the two Roman Catholic humanists saw the 'godless' work of the Rosicrucian at Massa's house in 1622, and wanted to reinstate

music, as it were, by creating a monument to it in its 'true' sense? And might the view of the banqueting-hall be a view of the atmosphere around Massa and Torrentius in 1622, perhaps even connected with Massa's wedding?

Matham's engraving could have been a step up to a tradition of detailed musical still-lifes with musical notes, as did take place in Italy after Caravaggio, especially in the circle at Bergamo around Baschenis; it was not such a step, but from that instant, music joined the current emblems of transiency and finity, joined the skull, the hour-glass, the watch, the portrait in the Vanitas painting, the smoking candle or wick, the tobacco and the pipe, the beautiful ripe fruit, and so many others.

Music in the 'Vanitas' is usually found in conjunction with emblems which in their turn belong to the vita voluptuosa, a life of delight: the accoutrements of eating, drinking, smoking, love-emblems; but also often in conjunction with emblems of the vita contemplativa, the contemplative life, such as books (usually old, faded, curled-up papers and parchments), death-emblems; and even occasionally with those of the vita activa, the life of action, such as battle-emblems like weapons, armour, trophies, a Vanitas by Leonard Bramer from about 1635[78]) being an example.

Recognisable Netherlands musical compositions, and those where the music itself indicates the message, were seldom found again in the 'Vanitas'. The fact that the picture showed that music was concerned was enough for the emblematism, and most painters confined themselves to that. There were however numerous in-between stages, from purely emblematically intended 'music' up to the concrete piece of music: pictural suggestion, 'any old notes', existing music

Leonard Bramer, Vanity, about 1635 Vienna, Kunst-
historisches Museum

reproduced fragmentarily and thus rendered unrecognisable or unperformable. In view of the fact that these are general characteristics occurring in other allegorical types too, in the paintings of gatherings and family portraits, and in genre-pieces, we shall return to this at the end.

Let us just take a look now at a few 'Vanitas' paintings with recognisable musical compositions or inferences.

It was to be expected that the Leiden School of 'fijnschilders' ('Feinmaler') would do justice to musical notes. However, little can be identified, for reasons which will be explained in the last section. In special cases they assumed the language of the notes into the emblematism. Their leader, Gerard Dou, took the lead here. A striking example is one of his late works (1667)

in the Lakenhal. The open music-book shows lightly-painted notes and staves, which do not at a first glance invite closer examination; they bear much resemblance to the vague, empty indications which so often occur. It is however well worth the trouble of taking a sharp look at this still-life of Dou's, which he executed in what was for him a modern technique. The barely visible letters then disclose the title: 'Memento Mori'; the music, in the tenor key and *imperfect*, alla-*breve* time, contains a series of intervals which are continually repeated: D-B-flat, B-flat-D, in various rhythmic combinations, joined a little later by an F. This leaves us with one conclusion: that it is a very suggestive imitation of a trudging funeral music — three heavy bells, do-mi-sol, toll their gloomy chimes out over the town of Leiden:

Edward Collier, also from Leiden, painted a Vanitas in 1684 which is also a self-portrait with a number of familiar Vanitas requisites. The work is a last offspring of the family tree of a Leyden Vanitas genre which was started by David Bailly — particularly with his famous work of 1651, now in the Lakenhal — who in his turn owed this aspect of his work[79]) to Dou. The painter is holding a drawing of a young woman: he had been widowed twice as a young man, so that the portrait is probably one of his deceased wives, or the one who was alive at the time, portrayed to-

gether with himself to intimate the brevity of life for both of them, as seen on the piece of paper sticking out of a book: 'Vita brevis ars longa' ('A short life is a long art'). Besides the skull, terrestrial globe, wick, watch and so forth, there are once more the wine-glass, pipe and music of the 'vita voluptuosa'.

The open music-book shows: 'Tanneken — Jacob van Eyck', i.e. 'Onan of Tanneken, gebroocken van IACOB van Eyck' ('broken, with variations and diminutions, by Jacob van Eyck') being a page of the first part of Van Eyck's 'Der Fluyten Lusthof', which

Gerard Dou, Vanitas - Still-life, 1667 (Leiden, Lakenhal). Photo Cor van Wanrooy.

Dou, Vanitas 1667, detail. Photo Cor van Wanrooy.

Evert Collier, Vanitas 1684 (Private Collection.
With the kind permission of Verlag Bruckmann,
München)

Casper Netscher, the violoncellist, about 1670
(Cracaw, State Collections of Art)

appeared in 1646: one flute lies on top of it, another is next to the skull! It is evident that the word 'Lusthof' (pleasure-garden) also was meant to express the idea of transiency, as a vain, worldly pleasure, and this is proved more conclusively by another similar Vanitas still-life by Collier[80]) in which there is an open copy of a song-book probably published by Michiel de Groot about 1678 and entitled 'Cupidoos Lusthof, ofte Amsterdamze Somer-vreugt' ('Cupido's Pleasure-garden, or the Summer Joys of Amsterdam')[81]). The only known example of 'Cupidoos Lust-hoff, ofte Nieuwe Uytrecht-se nachtegaal, singende d'aldernieuste Franse Voysiens en Amoureuse Gesangen . . . t Amsterdam . . . Michiel de Groot, 1679' is in the British Museum. The title of the book in Collier's Vanitas is, as far as is legible: 'Cupidoos Lust-Hoff, Amsterdamze Somer-Vreugt, bestaende in Verschyde nieuwe Voysen, Minne-lietjes . . . Door de vermarste Liefhebbers . . . t Amsterdam'. This obviously recalls the printer on the Nieuwendijk 'In de Groote Bijbel', Michiel de Groot, who died in October 1680, aged 46. And even if Collier invented the title, which is unlikely, he did mean a song-book. Both 'Lust-hoven' confirm in their turn the voluptuous conception of music.

Nothing further is known about the song 'Tanneken'. The title indicates a girl's name and most likely announces an amourous content. The other name in the title, 'Onan', would appear to indicate an obscenity, as was very usual in those days; but it could also possibly be a bastardisation of an older title. Van Eyck's music was still used forty years later . . . or the book is perhaps meant to express past glory in the painting, together with other 'old' works!

Finally, another Vanitas representation from the same period[82]). It is an allegory, for-merly ascribed to Jan Verkolje, now to Netscher, and possibly a real portrait: a richly-clad youth, still a milksop, playing the gamba. On his left are a musicbook and a viol on a magnificent Persian rug. Behind this the top part of a harpsichord in the Ruckers style can be seen, with the words which are particularly emphatic here: 'sic transit gloria mundi'. Youth, wealth and music — these glorious things disappear; but even the apparently indestructible, sturdy, fine architecture, the temple of art à la De Lairesse, which forms the background, will pass away too.

IV The Allegory of the Five Senses

Like Plato, humanistic morality pointed out the deceitfulness of the senses. As early as the sixteenth century we can see music pre-eminently involved with the Sense of Hearing. Particularly the Flemish artists and engravers had much influence in western Europe. Some representations accommodate the Five Senses all together, others show them in five separate versions, or in groups of two or three. We usually see a young woman, representing Musica — a motive borrowed from Antiquity — with a lute, and frequently with other instruments and an open music-book in which, incidentally, little or no musical details can be seen[83]). Of course the subject of 'good' and 'evil' music belongs here too. As might be expected, the Haarlem masters were connected with the Flemings here.

In 1617-1618 the most important Antwerp master besides Rubens at that time, Jan Brueghel, made for the Archduke, whose court painter he was, some monumental series of works on the Five Senses: one in five separate panels[84]), one in two works with two and three Senses respectively[85])

Jan Brueghel the Elder, The Sense of Hearing, 1617-
1618 (Madrid, Prado)

73

Jan Brueghel, Hearing, detail

Jan Brueghel the Elder, Taste, hearing and feeling
(Madrid, Prado)

In both series he depicted musical compositions for Hearing, and both series ended up in the Prado.

In the chief series, the one in five separate panels, Hearing is especially richly endowed. To the left of Musica, who, naked and beautiful, sits in the centre with a lute, there is a panoply of instruments and music. At the extreme left: a large harpsichord with two keyboards, decorated in the Ruckers style, confirmed by the just visible letters '-BANT', the end of the Ruckers motto 'acta virum probant'. The Vanitas idea peeps round the corner here. On the seven musicstands placed in a circle there are partbooks: one has on its title-leaf 'DI PIETRO PHILIPPI / INGLESE / ORGANISTA / DELLI SERENISS / PRINCIPI ALBERTO ET /ISABELLA ARCHIDUCHI / D'AUSTRIA ETC /DE MADRIGALI A SEI VOCI / . . .' (i.e. the second book of madrigals for six voices, dedicated by the Court organist Peter Philips to the Archdukes; the second printing came out in 1615). The other six musicstands of course carry the books with the six parts. The painter dedicated his work to the Archdukes by way of Peter Philips' visible dedication. In the background, on the left we see a company which performs the music. At the front, on the ground, semi-concealed by a Lira da braccio, are two canons relating to Hearing: the top one has the text 'auditui meo dabis gaudium et laetitiam' ('give my Hearing joy and gladness'; from verse 8 of Psalm 51, and the bottom one: 'Beati qui audiunt verbum Dei et custodiunt illud' ('blessed are they that hear the word of God, and keep it') (Luke XI, 28). The compositions are unidentified. Possibly they are by Phillips, who dedicated his 'Deliciae Sacrae' to the archduke in 1616[86]). There are two more 'Hearing' symbols in the musical corner of this painting: on the

table in the middle of the musicstands is an elaborate chiming clock; it announces to the ear the finiteness of time. There is a statuette in the base of this clock of a woman with a cross; she is Faith, and Paul says to the Romans (X, 17) and Galatians (III, 2 and 5): 'fides est ex auditu' ('faith cometh by hearing').

The other Jan Brueghel work referring to Hearing is also in the Prado[87]) and has the same canon on 'auditui meo' as on the panel described above.

The allegory of the Five Senses was more popular in the South (Theodoor Rombouts) and particularly in France than in the North Netherlands. In this connection I refer to Baugin's still-life in the Louvre from ca. 1630 with a lute and a French tablature book, and also to the one by Linard in the Benedict Collection (1627). A company of three ladies, a young gentleman and a negro servant by the North Netherlander Adriaen Hanneman, at Brunswick, is incorrectly noted in the catalogue as 'Allegory of the Five Senses'. The young lady on the left is holding a sheet of music with the bass part of a French amorous song belonging to the lute which the young man is tuning.

V Musica as Originator of Fortune and Love

Music doth heal — it is salutary, beneficient and restorative.

Music unites too — it creates harmony, agreement, love.

These related ancient ideas were probably nowhere as fertile in their re-birth as in the painting of the Netherlands. All the strings between strict morality and abandoned eroticism are plucked in the Portrait, the Family Portrait, the Company, the Loving Couple and the Genre of the North Netherlands. In ountless paintings and engravings, music was an extremely important, if not dominating function because of its musicians and dancers, instruments and music-books. There are infinite variations used by painters in both aspects separately or overlapping, or also in connection with the Vanitas and other allegories.

Music the healer was illustrated more in the South Netherlands, frequently with clarifying texts: captions to engravings, mottos on harpsichords. It was these very Ruckers instruments and those of other Antwerp instrument-builders which took the idea to the North. These texts, which go back to Antiquity via mediaeval theoreticians (Guido di Arezzo, Johannes van Afflighem, for example) and late-classical ones (Boethius, for example), can also be found in Vermeer's work.

In the North, music was much more important as a means to harmony and unity. The antique ideal and object of consideration was harmony in the individual, the family, society, expressed in and by music, to which the Platonic Eros confined itself because of moral considerations; this ideal widened its scope in an erotic sense in and after the Renaissance. This was particularly expressed in the art of the North Netherlands.

Where Antiquity had been limited to the practice of 'good' love and music and the condemnation of the 'bad', these regions exhibited an unhidden preference for the 'bad' and were not ashamed to practice and enjoy it. But here, too, there was a scale of love-emotions ranging from high to low. Our painters who involved music in their work presented us with a magnificent spectacle. In the work of Vermeer, who was the greatest painter of this type of subject, we see eroticism presented in the most ex-quisite fashion in that wonderful muted light; aristocracy prevails in the work of Thomas de Keyser and Ter Borch, a fine chiaroscuro in that of Ochtervelt, intimacy in that of De Hoogh, charm in that of Metsu, swagger in that of Van Mieris and Netscher, but intentional triviality, full of realism, is presented to us by Miense Molenaar and Codde, more so by the Van Ostades, culminating magnificently and yet somewhat tarnishing, and gaining equilibrium at the other extreme in the work of the brilliant Steen and the dramatic Brouwer. The best works of Schalcken, Jan Verkolje and Van Slingelandt also belong here. The secret but clear preference for 'bad' music and love in their domestic orgies, in the inns and brothels begins to take form with the depicting of music in bourgeois circles: one clearly senses the element of *insinuation*. It can be felt everywhere — in dancing (for example Codde), a plainly erotic intensification of the music, in the many music-books, open or closed, spread out on the floor, table or elsewhere. We find the entire gamut of the experiences and perils of love in connection with music: the good marriage and calm happiness in aristocratic circles (Thomas de Keyser in the Rouen museum), seduction, the threat of an ending love-relationship, of a rival. All these results of that powerful field of tension are interpreted with such conviction and intensity as to explain to an important extent the attraction which the paintings of the seventeenth-century North Netherlands have always exerted.

If painting — just as drama and literature and music too — was a double-faced mirror from antiquity right up to the eighteenth century,

i.e. a *speculum virtutis*, a mirror of virtue, a moralising image preaching how life real-

Abraham van den Tempel, Portrait of the family
David Leeuw, 1671 (Amsterdam, Rijksmuseum)

y ought to be — which mirror should be the real one, the 'per essentiam', the only one, —

and a *speculum vitae*, a reflection of everyday life as it really is, frequently in contradiction with the ideal, which ought actually only to be a mirror 'per accidens' which might be translated as 'by misadventure'!), and which is usually an offence to the speculum virtutis,

— then the North Netherlanders, with their sense of reality, usually chose without compunction the second face of the mirror, the reverse side', the *speculum vitae*.

It is obvious that a product as cultivated as a musical composition occurred more often in paintings of aristocratic and bourgeois circles. We shall give three examples of the various sectors.

In 1671, Abraham van den Tempel painted the wealthy Amsterdam merchant, David Leeuw, and his family[88]). Both the painter and the Leeuw family were Mennonites, rep-

resenting in this period a more austere group in Netherlands protestantism. Their sober garments and religious devoutness attempt to express this in the painting, but cannot conceal the ripe baroque splendour and wealth. The father presents his family to the spectator, the mother, Cornelia Hooft, has her youngest daughter, Susanna, almost two years old, on her knee, and the other four children are making music. They are about to perform a religious song in two parts with basso continuo: the eight-year-old Cornelia has the soprano part-book, eighteen-year-old Maria the contalto, Weyntje, nearly twelve, sits at the harpsichord and the only son, the fourteen-year-old Pieter, is about to play the viola da gamba. The bass part is on the floor next to him. The death of this handsome boy only six years later accentuates how realistic the idea of transiency was in those days. Only the melody ('canto 1') can be followed properly — that was the intention of the commissioner and the painter:

Wat is van smen-schen le-ven, Dat God elck heeft ge - ge-ven
Als 't niet werd daer be - ne-ven Door Go-des Geest ge- dre-ven,

En ver - he-ven om e - ven te stre-ven Nae 't eeu-wigh le - - ven.

The music is that of 'Lo Spensierato' ('The Thoughtless One'), the sixth of the seven Balletti a tre voci' by Gian Giacomo Gastoldi, published originally at Venice in 1594. They were famous throughout Europe, even more than his 'Balletti a cinque voci', and were published time and time again, sometimes in revised forms. One of these revised editions appeared in 1628 at Amsterdam, published by Willem Janszoon Wijngaert,

with a fourth part (ad libitum) by Godfried Oldenraet, an organist from Zutphen. This version had a sacred text by a certain Venerable gentleman V., for the purpose of edifying music-making in the home. Some time later this text seemed old-fashioned, although Gastoldi's music was still very much alive, and so Paulus Matthijszoon published a new edition in about 1650 with a fourth part by Jacobus Haffner, who was

Bartholomeus van der Helst, Musician, 1662 (New
York, Metropolitan Museum of Art)

employed in 1648 as organist of the Lutheran Church. As well as the old text, this edition contained a new sacred text, which is reproduced here. Although the music was not of Dutch origin, it had become part of the culture of the Netherlands.

This Sunday picture of the Leeuw family is a purely Dutch translation of the Antique 'harmony in the family', permeated still by the Vanitas idea, a Calvinist atmosphere and a certain degree of obligatory aristocratic snobbery. The idea of using music to show a 'good, harmonious family' is strongly emphasised. The harpsichord demonstrates the extent to which the family is filled with the conviction of worldly vanity: it has the words 'acta virum probant' on it, thus belonging to the 'sic transit gloria mundi — soli Deo gloria' group. The words in the text 'om even' indicate 'equally', 'together', expressing once more the general harmony. In this painting the decorative little dog seems to unite two very different emblem traditions. The presence of the animal, a 'kooikertje' (an old Dutch spaniel breed) in a voluptuous attitude, reflects the lapdog occurring in exactly the same way in comparable situations in many Dutch paintings of the period, as a symbol of erotic love: this significance comes out in the painting in question very faintly, but unmistakably. As was frequently the case elsewhere, the animal, in this fainter sense, has become one of the requisites and a decoration of family life. Another emblem is presented clearly, although weakened. The position of the dog with the two young children indicates this: it is that of a watchdog that repels hostile forces, especially those which threaten the lives of young children, the spirits of sickness and death, by his barking. This goes together with the rattle which one of the child-

ren holds in its right hand. The music is connected with this too. The origin of this complex of death-averting emblems is in the earliest history of mankind. The choice of the music itself shows that the Leeuw family wanted to extend the moral of this symbolic complex: may the children be spared from a premature death, not for life on this earth itself, but to have the opportunity in life to prepare themselves for 'streven nae 't eeuwigh leven.' As far as the first sentence is concerned, one can compare the pairs of lovers by Pieter van Slingelandt in Dresden, with music, instruments (a violin and a flute) and the lapdog, and three similar pieces in the Mauritshuis: Metsu's 'Company of Musicians' (about which more later) and Frans van Mieris' 'De Schilder en zijn Vrouw' (The Painter and his Wife) and 'Herbergtafereel' (Scene at an Inn). In connection with the second sentence, Godert van Wedige's so-called 'Hausmusik' from 1616 in the Walraff-Richartz Museum at Cologne.[89])

Bartholomeus van der Helst was a celebrated portrait painter in higher circles. A delightful work of his is the young female musician from 1662, now at New York. It is not clear here whether it is a portrait or an allegory. Be that as it may, the work is rich in emblematism as a portrait, and a beautiful girl was model for the allegory. The latter case seems to conform more with the character of the painting. The painter wove several motives together. Before everything else, the young woman with the lute is the authentic representation of the Antique 'Musica' as she had taken form in sixteenth-century Flemish humanist circles. She is sitting in luxuriant surroundings and tuning the strings of the soul to euphonious harmony: she is tempting to love and earth-

Pieter van Slingelandt, Pair of Lovers (Dresden,
Staatliche Kunstsammlungen)

ly pleasure. But the threats are not missing: the dark clouds and sinister vase with river-gods.

She is not only Musica, guiding us to harmonious love, she is as herself, a young woman, lovable, Divine. Her 'rosebud mouth, her milk-white neck and bosom' show us what poets were singing about since the great Hooft in arcadian poetry far into the eighteenth century: the beautiful nymph, shepherdess or goddess. The music sings of this too. The tenor part is lying open. The painter intentionally lets the word 'Iris' be visible: the goddess of the rainbow, linking heaven and earth, who even as a divine gift from heaven arouses virile desires. The music opens with a pattern which since the sixteenth century had been characteristic for songs ('chansons') with imitation:

At the right there is the Superius part-book, and another one too, probably the Altus. Perhaps this composition will be recovered one day: much has been preserved, but much has also been lost. It possibly has something to do with the popular tune 'La Belle Iris', prescribed several times by Jan Luycken in his 'Duytse Lier' of 1671.

There is also a hidden message in the painting. Although we by no means understand everything about emblematism and must therefore be extremely cautious, I do feel that one can recognise an emblem that occurs somewhere else: the viola da gamba is waiting for the young man, who at the invitation of the girl, will play on it together with her on her theorbo (for that is what her instrument is — a bass lute) united with her in complete loving harmony, the tenor part which is shown.

Nicolaes Knüpfer, who had settled in the cathedral town of Utrecht, is one of the excellent painters who are little known or not at all. He taught Jan Steen, and the painting which we shall discuss here, now at Dresden[90]), is fine testimony of both facts. He was born in 1603 at Leipzig, but soon moved into Netherlands circles, already working with Abraham Bloemaert at Utrecht before 1630, and surely underwent the latter's Caravaggesque influence. He himself, however, chose a more Dutch direction in which he could realise a purely intuitive Dutch humour.

The picture is a portrait of a family in a fine, sturdy, sober Dutch interior; there is reason to believe that it is the Knüpfer family itself. Knüpfer married in 1640, so that the painting can be placed somewhere around 1645, which is confirmed by the style. Here too, family harmony is symbolised by their making music together. What they are singing is surprising: 'So d'ouden songen, so pipen de jongen' ('as the old ones sung, so whistle the young'), a favourite theme by Jan Steen, so that Steen's direct source is known[91]). Seeing that Steen regularly portrayed himself on paintings with this and other similar motives, this, together with the chronological agreement and extremely spontaneous and involved representation, is a further reason for thinking this to be a Knüpfer selfportrait.

And the humour with which this delightful little work is filled! On the table, which is covered with a fine cloth, as if on a platform, and on a cushion too, is the youngest member of the family as the radiating center,

Nicolaes Knüpfer, Portrait of the painter and his
family (Dresden, Gemälde-Galerie)

Nicolaes Knüpfer, Portrait of the painter and his
family (detail)

Nicolaes Knüpfer, The painter with his family and
the celestial music (Paris, Musée du Louvre)

piping out the highest part, lovingly sup-ported by her mother. The mother has re-moved the snow-white wrap and 'revealed' her, as it were, and this divine putto stands there confidently, like a little Venus in her dazzling childish nudity. How effect-ive are the gestures of the jovial time-beat-ing father and the scamp of a pointing brother: from the base of the table they point like the sides of a triangle to the luminous apex. The little brother is not merely pointing to show that it is his sister, but also, like Cupid, at the focal figure in this painting: the young goddess of Love, the fruit of what 'the old ones are singing'. And so we too are in a upturned world, the 'verkeerde wereld'. It is not so important that the notes cannot really be sung. The painting is a particularly charming transi-tion from the work of a master to the more bantering and coarser style of his pupil, Jan Steen, who must have studied with Knüpfer during these very years.

An interesting supplement to this musical self and family portrait is another painting of Knüpfer's in the Louvre entitled 'The Apparition of Saint Cecilia to a Young Married Couple'.[92]) It has an even more allegoric character, and is more related to the style of Rembrandt's pupils such as Gerbrand van den Eeckhoudt. The title ought to be: 'The painter, Nicolaes Knüpfer, and his family, choosing celestial rather than earthly music'. Indeed, we are irrefutably confronted with a self-family portrait again, albeit from shortly afterwards. The couple stands in the middle of a landscape. Whilst Bacchants perform a wild round dance a little further away on the right, celestial light is visible on the left above the hills. where Saint Cecilia can be seen accompa-nying at the organ a fat little Dutch che-rub standing on the hill and singing from a book. Winged putti, also playing and clearly inviting, stream swiftly towards them. The easily recognizable daughter, a little older now, strews a path of flowers towards heaven for her parents, a delight-ful little Flora. The painter himself, clad in an idealistic costume — a bit of a Roman emperor and a bit of an Arcadian shepherd — leaves no doubt as to his choice: with a generous gesture (compare the piece at Dresden!) he points to and invites his wife — who is just as calm and fair as in the Dresden painting — towards the higher spheres. Perhaps the singing cherub on the hill is their youngest daughter, the little Venus of the other painting, now dead and singing before God's crown, which the Paris work places still more in the atmosphere of the Vanitas and of an Allegory on per-petuity. Just as Knüpfer in his painting 'So d'ouden songhen . . .' leads the way to his pupil Steen, he does it in the other one, with Saint Cecilia, to his other skilled pupil, the painter-organist referred to earlier, Ary de Voys.

Closing Remarks

Where motets appear in engravings and paintings in the South Netherlands, music has, besides its own symbolic meaning of ultimately eternally singing praise before the throne of God, no new symbolic significance because of its appearance in a painting or engraving: its appearance has no other function than to be executed, and so it is a remarkable kind of edition with at the most a votive or devotional character — a very fine specimen, indeed, of late manneristic and early baroque counter-reformational art.

Whenever we find a musical composition on a painting from the North Netherlands, there is definitely a double emblematism. The very presence of a musical element (an instrument, music-book, notation) is sufficient for music's general and emblematic role in visual art. If the musical element is a real composition, this composition has an additional special emblematism which is only involved in that particular painting, an emblematism with a more concrete significance. Between a vague impression of musical notes and the accurate reproduction of a composition, there is an area with many different manners of partial reproduction of music which are too interesting and too rich in meaning to neglect. The conception of this essay is such that I can only mention them briefly here — perhaps there will be an opportunity some other time to go into them in more detail.

Certain characters of painters would certainly never have copied or included musical compositions on canvas: the dynamically and monumentally expressive baroque masters (Rubens!) and the typically pictorial ones (Hals, De Claeuwe). The approved field of operation for portrayed music is that of the successors of miniature-painters: the 'fijnschilders'. But not all painters of this school who used music in their work painted performable music; and, put differently, it was not only these painters who painted musical compositions accurately, and the reason why some painters dispensed with making the music wholly recognisable does not have to be that they were not 'Feinmaler'! These, too, had to have a special reason for painting music.

It is fascinating to see how the Leiden School and related painters behaved in this respect. Gerard Dou and Frans van Mieris were masters of how to make it appear that a piece of music was reproduced in every detail — but if we peer more closely we are amazed at the masterful disguise which might be termed a kind of trompe-l'oeil. But occasionally a single word, or the type of print, tells us that the painter had a real book of music in front of him. This was usually an Italian book of madrigals, which can be seen by one or more italian words from the text, or by the indication of the part: e.g. 'BASSO'. There are marvellous examples in Dresden of works by Dou ('An Artist in his Studio') and in Schwerin by Van Mieris.

This method was also used by many still-life and genre painters: Peschier in his three well-known still-lifes. Thomas de Keyser in his wonderful painting at Rouen, which we have mentioned before, Bochout and others. I particularly refer to a somewhat untidily composed, but very interesting Vanitas in the Louvre by the Haarlem painter Vincent Laurensz van der Vinne, painted around 1650-'52 to commemorate the tragic death of Charles I of England. Here the madrigalesque music, painted 'trompe-l'oeil', belongs together with the hunting attributes and the precious shell to the 'vain' and 'Arcadian' delights which came to an end so suddenly and unexpectedly for the king:

Gerard Dou, The Artist in his studio (Dresden,
Staatliche Kunstsammlungen)

Gerard Dou, The Artist in his studio (detail)

Frans van Mieris, A couple making music (Schwerin, Staatliches Museum)

Frans van Mieris, A couple making music (detail)

Dutch anonymous, Allegory on the death of Charles I, King of England (Amsterdam, Rijksmuseum)

Vincent Laurensz van der Vinne, Vanity (Paris, Musée du Louvre)

'Denckt op t'ent' — be mindful of the end! The picture is not the only one of its kind, such paintings being quite a fashion in Holland at the time.

In these works, too, the italian madrigals and dances can be recognised by a small indication and as such emblematically as a love-song.

The very opposite also found its masters. Metsu's 'Company of Musicians' at the Mauritshuis shows a miracle of 'detail-painting' in music. The music appears to be only vaguely indicated. But if we take a magnifying glass to the panel, the sheet of music which the young woman has just written and the result of which she is appraising whilst beating time, reveals itself as a dainty real notation, the book to left of the lute-player turning out to be a real lute tablature.

In Vermeer's work we can see a different intention and working-method. In the works in which he painted music, it belongs together with other objects, such as the 'painting within a painting', to a kind of second zone: in their masterly outline it indicates only the core of a particular emblem — any detail would spoil the effect.

A painter like Steen used no music notes whatsoever: his dynamic flourish could not permit the eye to pause, to be forced to examine music more closely. In one painting, 'The Morning Hour' (Buckingham Palace), there is in one corner an exquisite musical still-life: a lute and a sheet of music. However, the notes are in virtuoso pictorial brushwork. Steen painted this picture in Haarlem in 1663, when he was under the influence of the circle around Hals, as were so many others: Pieter Claes also exhibited this more pictorial manner in his later work, such as a still-life with music from 1657.

There is another group of painters who used a lot of music in their paintings: the Utrecht painters Honthorst, Ter Brugghen, Van Baburen, Van Bijlert and others. In this respect they differ greatly from their great example, Caravaggio. Whereas this latter, in spite of his baroque chiaroscuro, created as heritage of the Renaissance a classically balanced and clear relationship among his components, and painted the madrigal notes in the music-books with realistic distinctness and full of symbolic meaning, his Utrecht successors had a more outward, pictorial and bold effect in mind which would have been spoilt by emphasis on little details such as musical notes, which were therefore avoided.

A musical composition in a painting, then, is more appropriate to the contemplative mood, such as that of a still-life, and especially that of the Vanitas. In the same way as painted words and textquotations, they intrigue the eye, attracting it as though to the focal point of a work, as though to a key to the mystery. They gave the painter and still give us today a sensation of something concrete in the imaginary world of the painting, and also frequently have the effect of optical illusions.

In the art of the Netherlands of the sixteenth and seventeenth centuries, these musical notes, these 'foreign bodies', led to a number of genres in painting, the Antwerp motet engravings and the Vanitas of the North Netherlands forming the peak. This was possible because they were part of an art which was such an essential element of humanistic painting: the art of music.

'Tempus et aeternum — haec meditanda homini':

'Time and eternity — on these things must man meditate.'

Gabriel Metsu, Company of Musicians (The Hague,
Mauritshuis). Photo A. Dingjan.

Caspar Netscher, The viol lesson (Paris, Musée du Louvre)

And they have been meditated upon, from all aspects, from inside and outside. The task was fairly shared by the North and South Netherlands. At the end of the seventeenth century the task was accomplished:

Vanitas and Allegory disappeared, and with them a culture. And we cannot do better than to close here with an eye-witness, Jan Luyken, who in his 'Spiegel van het Menselijk Bedrijf' in 1691, put it all in poetry, in his 'Instrumentmaaker' and 'Musikant':

De Instrumentmaaker.
'tJs goed, of quaad, Naa 't oogwit staat.

De Musikant.
Js 't Dropie soet,   Staat naa de Vloed.

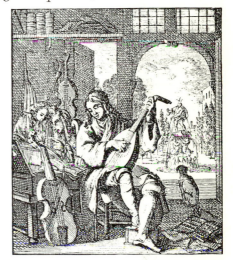

Het Snaarenspel, een spys der ooren,
Hoe aangenaam en uitverkooren,
 Dat van de Speelkunst werd bereid;
Soo 't uw Gemoed niet diend tot trappen,
Om tot den Oorspronck op te stappen,
 Dan is 't; als alles: Eidelheid.

Het Maatgesang en Spel der menscn,
Soo lieflyk als men ooit konwensen,
 Js maar een Staaltie van 't Geluid,
Dat opgaat uit der Eng'len Kooren,
Voor d' Eeuwyge Oorspronck van het hooren,
 Daar Vreugde nooid een Einde sluit.

From: Jan Luyken, 'Spiegel van het Menselijk Bedrijf'. 1691

How *pleasant* and *elect* is the *play of strings*, food for *the ears* prepared by the art of music; if it does not serve to the mind *as steps* to find the *cause*, it will be, like all, *vanity.*

However *charming* it is to *sing* and *play*, it is merely a specimen of the *sound* that arises from the *choirs of angels* for the *eternal origin* of *Hearing,* where joy knows *no end.*

Family Tree of Musical Conceptions in Antiquity and Humanism

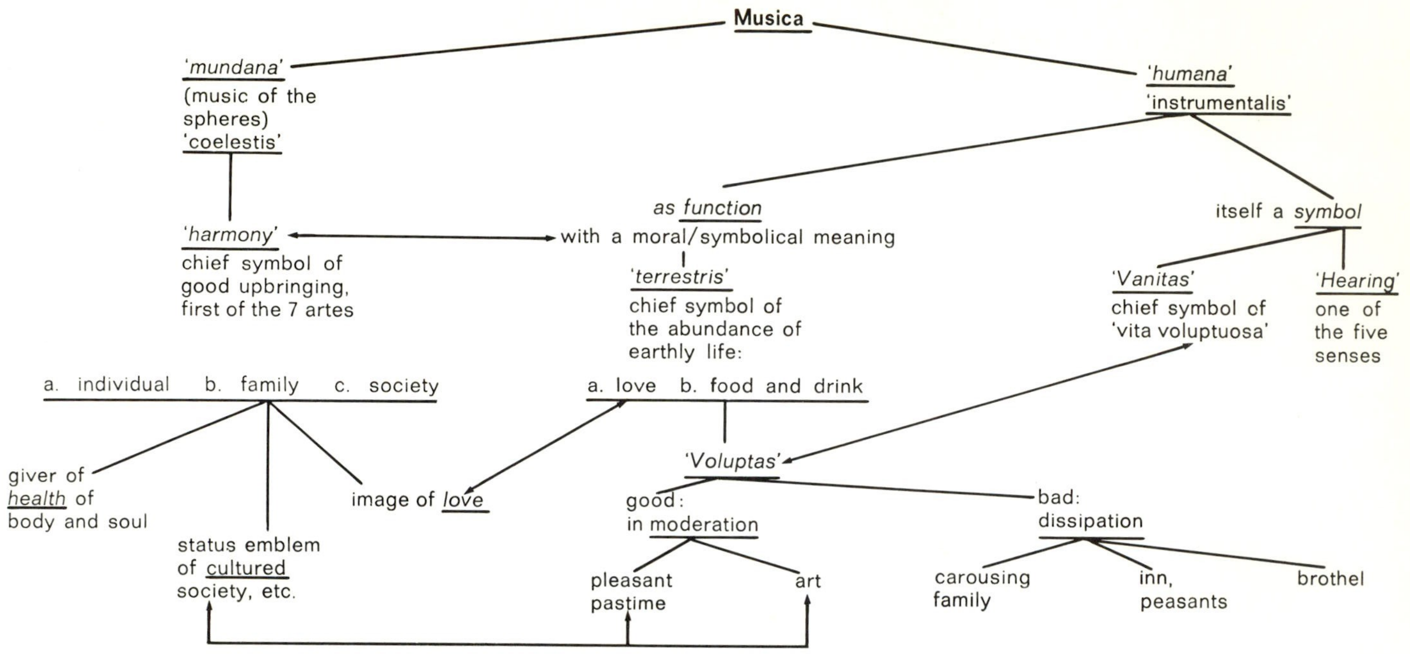

Musica

'mundana' (music of the spheres) 'coelestis'

'humana' 'instrumentalis'

'harmony' chief symbol of good upbringing, first of the 7 artes

as function with a moral/symbolical meaning

itself a *symbol*

'Hearing' one of the five senses

'Vanitas' chief symbol of 'vita voluptuosa'

'terrestris' chief symbol of the abundance of earthly life:

a. love   b. food and drink

'Voluptas'

bad: dissipation

carousing family

inn, peasants

brothel

good: in moderation

pleasant pastime

art

image of *love*

a. individual   b. family   c. society

giver of *health* of body and soul

status emblem of *cultured* society, etc.

Christianity with its parallel distribution, partly derived from that of Antiquity,

1) the music of the heavens and the angels.

2) the morally conditioned admissibility of earthly music, which is 'Donum Dei', permeates this system and its development.

Cosroë Dusi's painting over the altarpiece provided it with a nineteenth-century atmosphere of sweet romanticism.
See J. Lauts, Vittore Carpaccio (London 1962) 239, catalogue no. 39, fig. 193.

A rare example is a fresco from the Imperial Villa at Pompeii from ca. 60 A.D., a picture of two musicians, a remarkable parallel to the Dutch paintings with two musicians: one with a stringed instrument (cittern and lute respectively), the other singing and beating time with a music-book in the hand; it is not clear whether the signs on the Pompeii fresco are meant to suggest music or just words. See illustration in Geschichte und Gegenwart 11, table 26, fig. 2.

See H. Leichtentritt, Händel (1924), 22.
Roubillac died on January 11th 1762; the monument was placed in the Abbey after his death. It was his 'swan song'. Illustration: Musik in Geschichte und Gegenwart 5, table 55 (Opposite col. 1279/80). See Kath. Esdaile, The Life and Works of L.-Fr. Roubillac (London 1928) 154-156; Terence Hodgkinson in Victoria and Albert Museum Bulletin I (1965) no. 4, p. 1-13. For more symbolism in Händel's memorial, e.g. regarding the composer's 'fine ear for music', see Esdaile, op.cit. 155-156.

fol. 48r.

fol. 52r.

Prado.

Edinburgh, National Gallery (on loan from Holyrood Castle).

This painting, probably in a private British collection today, is reproduced in H. Besseler, Die Mensuralnoten und Taktzeichen im 15. und 16. Jahrhundert, 2nd. ed. 1906, oppos. p. 200, and in A. Pirro, Histoire de la Musique de la Fin du XIVe siècle à la Fin du XVIe (1940), oppos. p. 80. There was a replica in an Italian collection, see Bollettino d'Arte 1924-'25, p. 158. For the attribution to the Master of the Embroidered Foliage, compare the 'Madonna with Child in a Garden', reproduced in, among others, Friedländer, Die Altniederländische Malerei, IV nos. 84 and 84b, Pl.LXIII, and in the exhibition catalogue 'Bloem en Tuin in de Vlaamse Kunst', Ghent 1960, nos. 172/3, Pl.24 and 23, the exhibition catalogue 'Vlaamse Anonymi', 1969.

a Washington, Nat. Gall.
See: E. Winternitz, Musical Instruments and their Symbolism in Western Art (London, 1967), p. 145/9, Pl.66 and 67.

Ince Hall.

Lille, Musée Municipal, Inv. no. P. 816.

Fragment of the picture to which Ilja Markx-Veldman drew my attention, and which she will include in a publication in due course.

12 Particularly: Politeia 398-399.

13 The abbreviation 'auth' on the picture probably alludes to this; or could it be another term for 'fecit'?

14 Reproduction of engraving and transcription of the motet in: M. Seiffert, Niederländische Bild-Motetten vom Ende des 16. Jahrhunderts, Organum, I. Reihe, no. 19-20 (vol. I and II), Kistner & Siegel, Leipzig (1929); vol. II, p. 4 ff Seiffert incorrectly titles the picture 'Maria und Elisabeth mit dem Jesus Kinde'.

15 M. Seiffert, in Archiv für Musikwissenschaft I (1919), 52.
A. Hyatt King, Four Hundred Years of Music Printing (London, Brit. Mus. 1964) Plate XIII.

16 Ant. Auda, La Musique de Liège (Schaerbeek 1930), 153-154.
Seiffert (Arch. f. M.W. (1919) 50, 51) incorrectly made the abbreviation 'D' into the 'truly Dutch' name of 'Dirk', which was adopted in the Dutch translation in the XXXIVth edition of the Vereen. voor Nederl. Muziekgesch. (1920); also by G. Kinsky, Geschichte der Musik in Bildern (Leipzig 1929); 152, 3; A. de la Fage's description in his Dipthérographie Musicale, 487, is inadequate.

17 Seiffert, Organum, I. Reihe, vol. II, p. 12 ff.

18 Louvre.

19 Seiffert, op. cit. vol. I, p. 12.

20 Seiffert, op. cit., vol. II, p. 8 ff.

21 Seiffert, op. cit., vol. I, p. 4 ff.; Haarlem, Episcopal Museum, Inv. nr. 392, the text however being here Psalm 116, with the same canon indication 'Duo in carne una'.

21a Acquisition Sept. 1972. Reproduced in: Bulletin Rijksmuseum XXI (1973), No. 1, p. 37, fig. 18. It is possible, however, that the painting at Haarlem is a copy of the original.

22 Seiffert, op. cit., vol. I, p. 8 ff.

23 See Jos de Klerk, Muzikale Speurtochten in Haarlemse Historie (Haarlem 1960), p 22.

24 See H. Betke. Die Kunst am Hofe der Pommerischen Herzöge (Berlin 1937); Jos de Klerk op. cit. p. 28-30; idem in Mens en Melodie XIV (1959), 259-261, with illustrations of the Erlangen drawing and the Rügenwald silver relief; Joachim Gerhardt, Pommern (Berlin 1958), pp 51, 79 and fig. 217.

25 Seiffert, in Archiv für Musikwissenschaft I (1919) 63.
Ex. formerly in Bückeburg, Fürstl. Inst.; Paris, Bibl. Nat., Cabinet des Estampes.
Seiffert incorrectly: Ciapellius.

26 op. cit. 16.

27 See: Edm. Bruwaert, La Vie et les Oeuvres de Philippe Thomassin, (Troyes/Paris 1915), p. 68 and cat. no. 327, p. 92.
The great engraver Philippe Thomassin (1562-1622), the teacher of Callot, had already collaborated with Ciampelli (1577 Florence - 1642 Rome) and Soriano (1549 Soriano - 1620 Rome); in 1609 for the frontispiece with the portrait of Pope Paul V for the Missarum Liber I by Soriano, which the composer dedicated to the Pope, and which was published by Robletti at Rome.
See Bruwaert op. cit., cat. no. 286. Ex. in the Bibl. Corsiniana at Rome. This writer refers to Soriano on p. 87 as 'organist' of St. Peter's: not only was Soriano Maestro di Capella, but since 1608 no less a person than Girolamo Frescobaldi was organist at this principal Roman Catholic church.

28 Seiffert, Arch. f. M.W. I (1919), 64.

29 Van Mander referred to this early talent but also to his early death (Schilder-Boek, 1604).

30 As far as Gheyn's engraving is concerned: Seiffert, Org. 1. Reihe 20, vol. II, p. 16 ff. ex. Amsterdam University Library, The Hague, Gem. Mus.
MGG 12 col. 334, incorrectly states the converse contribution of the two artists: Kupferstich von J. D. Gheyn nach Za. Dolendo.
As far as De Voys' painting is concerned: see A. P. de Mirimonde, 'Sainte-Cécile. Métamorphoses d'un thème musical' (Genève 1974), p. 35, Pl.21 Christie auction, July 12th 1937, No. 142; the present location is unknown. Arn. Houbraken, 'De Groote Schouwburgh der Nederlansche Konstschilders en schilderessen' III (Amsterdam 1721), p. 162-163; the incorrect birth-year 1641 (also in De Mirimonde and others) was stated by him.

31 Hollstein 787, act. 'Zuid-Nederlandse Grafiek uit de zestiende Eeuw' Museum Boymans - Van Beuningen, 1965, no. 112.

32 See G. Reese, Music in the Renaissance (New York, 2 1950), 393-394.

33 See Fl. van Duyse, Het oude Nederlandse Lied III ('s-Gravenhage-Antwerp 1907), 2574; R. Lenaerts, Het Nederlands Polifonies Lied in de zestiende Eeuw (Mechelen-Amsterdam 1933), 51.

34 Fl. van Duyse, op. cit. III, p. 1933; only the Bassus has been preserved of this edition, at Brussels, Koninkl. Bibl.

35 Reese, op. cit.

36 cat. no. 3.

37 For the Seven Hour collegia in Amsterdam and other Dutch towns, see: Jaarboek Amstelodamum XXVII (1930), p. 29/42, where however far too late a date is suggested for the earliest polyphony in

Amsterdam. For the ones at Leyden, particularly: A. F. J. Annegarn, Floris en Cornelis Schuyt (Utrecht 1973), p. 13/18. For Haarlem: Jos de Klerk, Haarlems Muziekleven in de loop der tijden (Haarlem 1965), p. 7/23.

38 Cat. no. 523-525.
See illustration in G. J. Hoogewerf, De Noord-Nederlandse Schilderkunst III (The Hague 1939), p. 112, fig. 55.
J. F. M. Sterck, De Heilige Stede in de Geschiedenis van Amsterdam, (Hilversum 1938), opposite 120 (ascribed there to Jan van Hout).

38a Because of its small size, the painter was not able to reproduce on this sheet of music, the words the Angels are singing. Therefore he has made them descend towards Maria, queen of the Angels, solidified in golden letters by the goldsmith in the golden crown with the twelve precious stones on her head: ,,Gloria in excelsis Deo''!

39 See Sterck, op. cit., 99; and idem. Van Rederijkerskamer tot Muiderkring (Amsterdam 1928), 51.

40 Martin Picker, in Journal of the American Musicological Society XII (1959), 94-95; XVII (1964), 134-143.

41 See E. de Jongh, 'Realisme en Schijnrealisme in de Hollandse Schilderkunst' in the exhibition catalogue ,,Rembrandt en Zijn tijd'', Brussels 1971, pp. 150-152.

42 Amsterdam, municipal archives, Thesauriers bills, 'Extraordinaire Saecke' 1604; B. v. d. Sigtenhorst-Meyer, J. P. Sweelinck en zijn Instrumentale Muziek, 70.

43 Fine examples are to be found on existing instruments such as the one made in 1618, now at the Berlin Institut für Musikforschung, and the one in the Copenhagen Musikhistorisk Museum, formerly in the possession of Gustav Leonhart; furthermore on numerous paintings by Dutch seventeenth-century painters, more of which will crop up in this essay.

44 Illustration in Oud-Holland 1970, 147 in I. Borgstrøm's article, 'Jacques de Gheyn als Vanitas-Schilder'.

45 See J. C. van Regeteren Altena, 'The Drawings of Jacques de Gheyn' I, Amsterdam 1936: Appendix III;Borgstrøm. op. cit. was also struck by this passage.

46 Larson Collection, Stockholm; see cat. 'Vanity of Vanities', no. 12.

47 With regard to Bloemaert's painting: See catalogue 'IJdelheid der IJdelheden', Leyden 1967, p. (XIV) and No. 3, p. 3 - 4.
With regard to De Vos' painting: See also R. Wittkower, in Miscellanea Leo van Puyvelde (Bruxelles 1949), pp. 117-123. De Vos' panel was presented at Christie's, London, July 31st 1939, No. 181, as an Abraham Bosschaert.

According to Van Mander, he owned a 'Glass with Flowers' painted by Hieronymus Bosch!

Utrecht, Centraal Museum.

See A. Bredius, Künstler-Inventare IV, 1173-1177.

Het Schilder-Boeck, Haarlem 1604, fol. 293v.

Städelsches Kunstinstitut, Frankfurt am Main. Illustration in the revised edition of Van Mander's Schilderboeck (Wereldbibl. Amsterdam-Antwerp 1936, 4 1950), fig. VIII, after p. 16.

Inventory 1621; see cat. 'IJdelheid der IJdelheden', p. XIV.

The inauguration and appraisal of the organ most likely took place on May 24th or one of the days around that date. The date of 'June' in the Tijdschr. v. d. Vereen. voor Nederl. Muziekgesch. XXII-2 (1971). p. 130, is based on the payment to the innkeeper on June 25th at whose inn the organ appraisers ate together with the burgomaster; the date of the bill is that of St. Jan, the date of one of the quarterly instalments.

This Latin pun is my own; but it fits in so well with the line adopted by the humanism of that day with its similar puns, such as De Gheyn's 'humana vana', that one might assume that such an obvious combination was indeed made.

Karlsruhe, Staatliche Kunsthalle. Illustration in cat. 'IJdelheid der IJdelheden', no. 17.

See Tijdschr. v. d. Vereen. v. Nederl. Muziekgesch. XXII-2 (1972) pp. 95-96; see also the bibliography there.

See Thieme-Becker ad vocem 'Hauer'. The engraving is mentioned and described, but the iconographically most important component, the niche with the skull-vase, is omitted.

Amsterdam, Rijksmuseum, inv. no. 2311-D-1.

A. J. Rehorst, Torrentius (Rotterdam 1939).

See Rehorst, op. cit. 191.

Rehorst, op. cit. 18.

Autobiography in Kon. Bibl. at The Hague; published by J. A. Worp in Bijdr. en Meded. v. h. Hist. Genootschap XVIII (1897); translation in Oud-Holland IX (1891) 131; A. Bredius, Joh. Torrentius (1909), p. 4; A. H. Kan, De Jeugd van Const. Huygens (1946), p. 83.

1931, p. 42.

op. cit., 75.

E. de Jongh. Zinne- en Minnebeelden van de zeventiende eeuw (Amsterdam-Antwerp 1961).)
Idem, in Openbaar Kunstbezit 1967, 29.

67 J. L. A. A. M. van Ryckevorsel, 'J. T., Rozekruiser en Schilder' in Hist. Tijdschr. XIV (1935). A. J. Rehorst, op. cit., 166 et passim.

68 See also Catalogue of Paintings Rijksmuseum, 1960, 305.

69 See E. de Jongh, op. cit., note 77.

70 See Bredius, Künstler-Inventare IV, 1173 ff; J. Bruyn, In Oud Holland 66, (1951), 149-150.

71 J. v. d. Vondel, Het Lof der Zeevaert (Amsterdam, Willem Jansz Blaeuw, 1623) closing verses 477-478.

72 Rijksmuseum Amsterdam, inv. no. 1084.

73 Ex. in municipal archives, Haarlem.

74 I. Borgstrøm in cat. 'Vantiy of Vanities', p.x.

75 My thanks for the identification and the reference to De Vitry to Prof. Dragan Plamenac.

76 W. J. A. Jonckbloet and J. F. N. Land, in their 'Correspondence et Oeuvres Musicales de Constantin Huygens' (Leyde 1882, p. CXLV), make the unjust statement about Matham's musical reproduction: 'un peu maltraité par le graveur': perhaps they had a bad copy.
The play on the words 'casu, quae', by Jos de Klerk in his Haarlems Muziekleven in de Loop der Tijden (Haarlem 1965), 121, is of course incorrect.

77 Galleria Doria Pamfilj, no. 384.

78 Painting in Vienna, Bramer made an etching of it himself; see illustration in D. A. F. Scheurleer, Het Muziekleven van Amsterdam in de zeventiende Eeuw (The Hague, without year), p. 117.

79 Munich, An Art-dealer's Coll.; illustration in W. Bernt, Die Niederländischen Maler des 17. Jahrhunderts (Munich 1960), vol. I, no. 191.
For the work in question by Bailly, see exhibition catalogue 'IJdelheid der IJdelheden' no. 1; B. Haak, in: Antiek II (1968), p. 407/10; M. L. Wurfbain, in Openbaar Kunstbezit XIII (1969) No. 7.

80 Lakenhal, Leiden, Inv. no. 237. See cat. 'IJdelheid der IJdelheden', no. 9.

81 See text in the said catalogue on p. 8.

82 Cracow, Wawel Museum. Illustration in J. Bialostocki/M. Walicki, Europäische Malerei in Polnischen Sammlungen (Warsaw 1956), fig. X.

83 Compare Martijn de Vos' Terra.

84 Prado, nos. 1394-1398; Hearing is no. 1395.

85 Prado, nos. 1403 and 1404; these are contemporary copies on canvas of the originals, which were burnt in 1731.

86 S. Speth-Holterhoff, Les Peïntres Flamands de Cabinets d'Amateurs (Brussels 1957), says on p. 56. referring to note 72 on p. 206, that F. Clerici, Allegorie dei Sensi de Jan Brueghel (Florence 1946), p. 29. identified the music on the panel; this is incorrect: Clerici names a few well-known musical-theoretical works of the sixteenth century in general which might be expected to be in a library around 1617 - but which do not occur here!

87 No. 1404.

88 Rijksmuseum, Amsterdam, inv. no. 2289.

89 My thanks to Mrs. Aafke Teensma, Amsterdam, for identifying the melody as a composition of Gastoldi; and to Mr. H. Keyser, Amsterdam, for referring to the emblematism of the dog as an exorciser of evil spirits.

90 Gemälde-Galerie, cat. no. 1258.

91 This saying was first used by Flemish masters, by Jordaens for example in a painting from 1638, now in the Museum voor Schone Kunsten at Antwerp

92 Illustrated in: A. P. de Mirimonde, o.c., p. 34, Pl.20.

Index of Persons and Place Names

SWETS & ZEITLINGER - AMSTERDAM

A